AF292147

"AI's accelerating pace has exposed how ill-prepared our education systems are for a world of machines, having long failed to cultivate core human capacities such as critical thinking, agency, and creativity. In a moment of collective disorientation, Reclaiming Purpose offers an urgently needed, reassuringly humane, and eminently practical path forward."

—Sanjay Sarma
MIT professor of engineering,
Former President, Open Learning at MIT

"Reclaiming Purpose is essential reading for anyone seeking to understand, and take responsibility for, the AI revolution now transforming higher education. Readers familiar with Paul LeBlanc's earlier work and distinctive voice—his precise, elegant prose rich with meaningful anecdotes—will find this book especially rewarding, a work filled with wisdom, deep analysis, and grounded common sense."

—Arturo Cherbowski Lask
Executive Director, Santander Universidades

"Reclaiming Purpose is bold without the boosterism and hype animating the AI landscape today. The AI earthquake may be a technological phenomenon, but the vision of this important book is uniquely and unapologetically human. The breadth of the book's vision inspires, but it also leaves higher education leaders with clear steps institutions can start working on immediately to rebuild and reimagine education."

—John O'Brien
President and CEO, EDUCAUSE

"We urgently need intelligent, informed, and humane guidance in shaping a world that will be dramatically reshaped by AI. How can we avoid its perils? How might it become an engine of both individual and collective flourishing? Read this illuminating, very important book."

—Rick Weissbourd
Senior Lecturer, Harvard's Kennedy School of Government and Graduate School of Education, author of The Parents We Mean To Be: How Well-Intentioned Adults Undermine the Moral and Emotional Development of Children

"AI, AI everywhere, but not a drop of wisdom to drink. In a world saturated with AI, progress does not come from abundance alone. Reclaiming Purpose *challenges us to design a future where human creativity and wisdom lead."*

—Laura Ipsen
President and CEO, Ellucian

"This is a book that covers AI's impact on society, the evolution of humanity, and the unique demands now facing education. It is a must-read for every parent, student, policymaker, and educator. This book can serve as a foundation for a new model of education and human development in the Age of AI."

—Aaron Etingen
CEO, Global University Systems

"This book is a must-read for everyone responsible or interested in the future of higher education and the future of humanity. This is a voice and a vision every higher ed leader should heed. And for all of its sweeping thinking about the future, it also offers practical steps that college leaders can and should take right now to best prepare for what lies ahead."

—Chris Gabrieli
Chairman, Massachusetts Board of Higher Education,
CEO, Empower Schools

RECLAIMING PURPOSE

RECLAIMING PURPOSE

THE UNIVERSITY IN AN AI WORLD

PAUL LeBLANC

WITH

TANYA GAMBY AND **GEORGE SIEMENS**

WILEY

For Simon

Contents

AI Disclaimer

I used AI in the writing of this book, and because I believe that transparency in the use of AI is an ethical imperative, I will identify those uses:

- First, I used AI as a research assistant, sometimes when seeking an answer to a quick question, such as "Who were the leading researchers that attended the Dartmouth Conference of 1956," and sometimes for more complicated items, such as "explain *backpropagation*."

- Second, I used it as an editorial assistant, as when I asked it to review all nine chapters of the book and to flag redundancies. It was incredibly good at such a task and saved me a lot of time.

- Finally, I used it as a synthetic reader in role-play, uploading a chapter and asking, "What would a critic say of our argument here?" Again, it was a valuable partner in flagging gaps in my arguments or perspectives I had not considered. I then revised accordingly.

I did not use AI to write, as it remains "meh" at best and I write *to think*, a function I did not want to surrender to a large language model.

Foreword

For years, we have watched higher education's eighteenth- and nineteenth-century foundations strain against the weight of modern learners whose lives, expectations, and futures no longer match the world for which our institutions were built. As a result, public confidence has waned, governments have intervened where they have seen internal leadership wanting, and concerns about affordability and value have signaled to many students that college is not for them.

For more than a decade, I have worked alongside Paul LeBlanc to confront these issues. In reform efforts, policy initiatives, and our work at the American Council on Education (ACE), we have wrestled with how best to help higher education meet the needs of individuals and society. We have tried to be equally skeptical both of desires to return to "the good old days" (whenever those were) and of the overwrought promises of various innovations to usher in a new era without fuss or friction.

Reclaiming Purpose seeks to provide readers with a careful and inspiring view of how the purposeful use of artificial intelligence (AI) can simultaneously free higher education from the limiting structures of the recent past and restore ideas and challenges that have informed higher education for centuries. Simply put, it posits that AI can recenter and reinvigorate a system that has too often confused transmitting knowledge with transforming learners.

At the heart of this work is a deceptively simple insight: higher education's modernization will come not from doubling down on epistemology—what students must know—but from rediscovering ontology—who they can become. Knowledge matters. Skills matter. But neither matters as much, or as lastingly, as identity. Learners and their communities thrive not when we tell them what to memorize, but when we help them understand who they are, what they can do, and how to navigate toward where they want to go. This is the knowledge that tomorrow's citizens need to create kind and stable communities.

When nearly all knowledge is frighteningly accessible, its transfer becomes an almost trivial pursuit. Learners and teachers need to ask what we need to do with all these things we know. What are the positive goals to pursue? Which ones promise heartache and destruction? What should be the ethical and moral framework by which we live our lives and teach our children? How do we assess the validity of knowledge claims, whether they are made by humans or machines? What new knowledge should we pursue and why? Learners need institutions that can help them ask these questions.

For generations, we have shied away from these questions in higher education, fearing, with reason, the hegemonic power of a single ideology. The result has been, on the one hand, an openness to very different ethical and moral frameworks, but, on the other hand, a weakened ability to resolve the big questions of who we are in the world and to how to live in that world. This has not always nor everywhere been so.

In America's colonial colleges, the final course for undergraduates was on moral philosophy taught by their presidents. The idea was to challenge students as they moved into the world, to take stock of that knowledge, and forge a code for living. Today, faith-based institutions proudly carry on the tradition of moral formation. Through our work in ACE's Commission on

Faith-Based Colleges and Universities, we have engaged leaders to discuss how best to animate the search for meaning and value across higher education. Those outside the academy remind us that the multilayered experience of the residential college is fertile ground for thinking about and testing moral understandings and ethical dilemmas. The "thickness" of these experiences is fertile ground for curated conversations about virtue, identity, and belonging. If only we could plant those seeds and nurture them more intentionally.

For many critics, AI is the ultimate in dehumanization of learning and teaching. For LeBlanc, AI represents the opposite: the potential for more time, space, and energy devoted to humanizing learning and building capacity to engage the central questions of meaning, belonging, and purpose more fully. This is the breakthrough *Reclaiming Purpose* introduces and the hope it inspires.

In this context, AI becomes not a threat, but an opportunity. Properly understood and thoughtfully implemented, it is precisely the tool higher education has been missing. It can personalize learning at a depth no institution has ever been able to offer, making knowledge transmission more efficient and effective. It can guide learners through decisions that used to depend on luck, confidence, or access. It can make pathways visible, skills attainable, and support immediate. Most important, it can help us create environments where students learn not only information but also possibility.

This book does not romanticize AI, nor does it dismiss legitimate concerns. Instead, it presents a grounded, humane, and actionable vision for how AI can enable the ontological shift higher education so urgently needs. It invites institutions to rethink not just their technologies but also their purposes. And it offers readers—whether educators, administrators, or policymakers—a way to navigate change without losing sight of what makes education meaningful.

Most encouragingly, it gives direction to the real reform that will make our institutions servants of our students' and society's needs. Faculties, with a new relationship to knowledge transfer, can focus simultaneously on the creation of new knowledge and the ethical, moral, and social transformation of individual students in their care. As LeBlanc notes, we can all point to the one, two, maybe even three teachers who made this kind of difference in our lives. Positive AI can multiply that number to the benefit of individuals and society. Institutions, in turn, can and must change the way they hire, advance, and develop these skills in faculty themselves to make the shift from epistemology to ontology real. Students, for their part, will be challenged to work harder and in new ways that will prepare them for a dynamic world and for lives that nurture themselves and others.

I have watched these ideas develop, strengthen, and mature over years of work with LeBlanc. I have seen his commitment to students, his intellectual generosity, and his unwavering belief that higher education can and must evolve. This book is the culmination of that commitment.

I am honored to introduce it, and even more honored to invite you into the conversation it begins.

—Ted Mitchell
President, American Council on Education
Washington, DC

When the World Changed

Renting an apartment in Paris for part of the year, reading late into the evening, spending time with a soon-to-arrive grandchild, and more time with my wife, Pat, writing a new book—these were on my how-to-spend-my-sabbatical-year list of possibilities for when I would finally step down from the presidency of Southern New Hampshire University (SNHU) in June 2024. It had been a great twenty-one-year run working with my team to create the country's largest university, with two hundred and fifty thousand students enrolled, serving student populations too often neglected by traditional higher education, and inventing new models of education along the way. But it was time to assume the more contemplative and measured role of scholar, thinker, and writer.

Then November 2022 happened. OpenAI released Chat GPT to the public and the world changed overnight.

Like most people, my initial exploration of artificial intelligence (AI) for the masses began playfully. At a holiday dinner party, my laptop sat at the end of the table like a strange guest, and we asked ChatGPT to write about higher education in the form of a Shakespeare sonnet, then in the style of e.e. cummings, then with the voice of a 1930s gangster. We asked it to outline a three-day itinerary for an upcoming trip to New York, adding our preferences for museums, cuisine, and theater. Its responses were uncanny in their speed, responsiveness, and detail. In subsequent days, I asked it to design an Intro to Poetry course for me, providing parameters that included course outcomes, assignments, and activities. I asked it to teach me how to create pivot tables in an Excel spreadsheet. I uploaded a chapter from my book *Students First: Equity, Access, and Opportunity in Higher Education* and asked for a summary. In task after task, it performed exceedingly well. I was thunderstruck with the possibilities.

I had this experience with breakthrough technology before. In the 1980s, I was one of the first doctoral students at the University of Massachusetts to have my own word processor, made possible through a friend at Wang Laboratories and his employee discount. I stood with a foot on each side of a new generational divide between those who typed (or paid typists) to produce essays, bottles of Wite-Out at their side, and those who used a word processor and paid no price for retyping whole pages when revising or making simple mistakes. Just a few years earlier, I typed my master's thesis using onion skin paper (look it up, kids) and cursed my many typos and corrections. Then I wrote my dissertation on my new word processor, sliding 5.25 discs in and out of the computer to do even simple corrections (goodbye Wite-Out), and yet feeling like I was working at lightning speed. It was the graduate student equivalent of going from a horse and buggy to a Model T. However, the real revelation came through a new campus writing lab with Digital Equipment Corp.'s Rainbow personal computers linked through a local area network. Because the full-time writing faculty would not go near these new machines, they ordered lowly teaching assistants like me to figure out how to use them to teach writing.

Task assigned, I brought my students in, and together we learned to use the computers. The students loved working on them and were engaged in ways I was not seeing when they were writing longhand in my regular classroom. They were more willing to revise their work, a task made easy by the "delete" key and ability to cut and paste. They loved online discussions, and when I turned on the anonymization feature, removing names from posts, I watched students otherwise silent in class find their voice, resulting in more robust online discussions. I was blown away and soon changed the focus of my research from literature to the nascent field of computers and writing. It was the start of the computer revolution, and thinking the world would be invented anew, I wanted to be part of it.

We readily understand the ways new technologies dramatically alter our ways of *doing* things. For example, transportation technologies like railroads, cars, and airplanes radically changed travel and thus radically changed the ways we conducted trade, organized our lives, and remapped the landscape of human activity around the planet. In my research, I looked at the ways new technologies altered the *cognitive landscape* of societies, how we come to *think about* things. Because writing has been with us for so long and is something we learn to do so early on in life, we can forget that it is a kind of technology, a profoundly powerful one when it was introduced into ancient oral cultures with no scrolls, codices, or books available to reference when one could not remember something. Memorization was the single most important cognitive tool of a learned person, and what one remembered literally equaled what one knew. The arrival of writing did more than offer a new way to store information; it rearranged the mental landscape of entire civilizations. In a world of pure orality, knowledge lived in the body and the community—recited epics, genealogies held in memory, wisdom encoded in proverbs and proverbs encoded in rhythm. As Walter Ong observed, oral cultures thought in formulas and narrative because memory demanded it; thought must be portable and repeatable, wisdom was a living performance rather than a silent text.[1] Homer's *Iliad* and *Odyssey* are full of tropes, repetitions, and alliterative phrases because the poems were memorized and recited long before they were written down for posterity; those mnemonic devices simply made the lines easier to remember. To put knowledge on a surface—clay tablet, palm leaf, papyrus—was to liberate it from the fragile and *embodied* human brain and, paradoxically, distance it from lived context. The word is not the actual thing; it is an abstract signifier of the thing. Jack Goody called this transition "the domestication of the savage mind" not to disparage oral intelligence but to mark the way writing disciplines or transform thought into lists, categories, and abstract thought and analysis.[2] Whereas memory in an oral culture was one's most important cognitive power, it faded as the primary intellectual virtue, much to Plato's chagrin, who railed against the new technology of writing because people would lose the power to remember. Turns out he was right. Where would we rank memorization today in terms of important cognitive skills? Knowing became not *holding* information but knowing where, and increasingly how, to *find* and interrogate it.

If writing externalized memory, print industrialized consciousness. When Johannes Gutenberg's press made identical texts reproducible at scale, the world did not just get books but new habits of mind. Reading became private and silent, a solitary act of interiority rather than a communal performance. The Reformation, the scientific revolution, and democratic politics are difficult to imagine without that shift. Ideas gained an independent life in pages and libraries, and with that independence came a revolution in skepticism and inquiry. The scholar no longer debated memory against memory but book against book, edition against edition.

These transformations—writing's cool logic, print's rigorous reproducibility—are reminders that every communication technology quietly changes the way a society thinks, and as such they change the world. Literate people developed analytic habits because texts made analysis possible; print-rich societies think in sequences and citations because printed pages reward such thinking. Eric Havelock argued that the Greeks did not merely invent literacy; literacy helped invent a new kind of Greek mind.[3] It is a reminder of that old axiom that we shape our tools and then our tools shape us. And one wonders, looking at today's large language models (LLMs) and generative systems, whether we are again at such a threshold. Just as writing once freed thought from the breath, and print freed it from the scribe, AI may yet free cognition from the page itself, inviting us not only to ask *what* we know, but *how* we will think when knowledge is everywhere and memory is optional. The past is not a mirror of what is coming, but it is a warning: meaningful technologies rarely arrive quietly, and they never leave the mind unchanged.

More than four decades after studying the relationship of technology to cognition, those first interactions with ChatGPT once again made me feel that the world was about to change—only this time more profoundly. First, the speed of adoption is dizzying: if generative AI was truly revolutionary, the revolution arrived fast. ChatGPT reached one million users within five days of its release and one hundred million users within two months.[4] We have never seen a technology adopted with this speed.

Shortly after its release, I was in San Diego at the ASU+GSV Summit, the largest gathering of education technology entrepreneurs, investors, institutions, journalists, and companies in the country, a must-attend event for anyone interested in the intersection of technology and education, with over seven thousand attendees. I ran into my old friend and fellow Canadian George Siemens, a renowned researcher in learning science and analytics, widely credited with creating the first massive online open course, or MOOC. He had a joint faculty appointment at the University of Texas and the University of South Australia, and at the latter institution directed C3L, a world-class research lab full of data scientists and learning researchers. A prolific researcher and expert in AI and learning, Siemens has been cited over forty thousand times in academic journals and articles. He and his wife, Tanya Gamby, had just moved to Adelaide in South Australia. George and I sat outside in San Diego sipping coffee and sharing our sense that the world was utterly changed. That AI, around since the famous 1955 Dartmouth Conference at which the term was coined, finally had its catalytic moment with the arrival of ChatGPT. That everything—society, medicine, science, the creative arts, technology, and yes, education—would or at least *could* be reinvented in dramatically new ways. We sat and wondered aloud about the role of humans if and when we are no longer the most powerful cognitive entities on the planet. What would we educate people to do, to think, to know? Indeed, would we have to rethink knowledge itself, or at least our relationship to it? This would be big. Like, invention of fire big. Like, invention of electricity big.

As our coffees got cold and we warmed to the subject, I said, "George, this is not public yet, but I will be stepping down from my role at Southern New Hampshire University in June 2024. Why don't you leave your posts and join me, and we can reinvent learning together?" When I said it, we both paused, then simultaneously and nervously laughed out loud, a tacit commentary on the audacity of the goal (and the idea that two Canadian immigrants with humble roots might take on such a task). Even then, we were witnessing a rush to announce AI point solutions that might serve part of the learning ecosystem, such as AI tutoring and back-office functions like transfer credit evaluation. Companies that had been in the market a long time were quick to declare that they were "AI powered," and one founder told me, "We have always been using AI. We just did not talk about it before." George and I were uninspired by the sprint to announce AI-powered point solutions that might serve parts of the learning ecosystem, such as AI tutors and AI student support applications. Those solutions might be helpful (even if the claims sometimes felt a bit overheated), but we were asking bigger system questions and speculating on how we could use AI to *reinvent education* to improve it and make it available to the millions of people globally who did not have access.

We were inspired by the book *Power and Prediction: The Disruptive Economics of Artificial Intelligence*, which argues that the full potential of AI is only unleashed when we use it in system redesign rather than a point solution within an existing system.[5] We asked ourselves, "What would it look like if we could use AI to reinvent education wholly unencumbered by the way the existing system operates, giving no thought to accreditation or financial aid or traditional roles? What if we took a clean sheet of paper approach to the question, creating an AI-powered, human-centered model of education for the dawning Age of AI?" Our excitement at saying aloud what we were not hearing around us at the summit energized us, and then we shook on it and said, "Hell, let's do it." Goodbye Paris, that pile of books, any time to write. While we did not exactly know what we had just agreed to do together, I knew that long-awaited sabbatical would instead become a journey into the astonishing, puzzling, sometimes frightening world of AI and the opportunity to reinvent a model of education that felt to us increasingly broken, ill-suited for the new-world order, and out of reach of too many.

In the following months, George and I exchanged ideas and started to play with the ways in which AI might alter education. We started with existential questions and a growing conviction that the *epistemological* focus of education, centered on knowledge—what one should know and how we know—would shift to more existential and *ontological* questions of being, how we think about being human, in community, and in a world where we will coexist with knowledge machines far more powerful than we are. We also agreed that genuine personalization of learning is possible, a goal long promised and always unrealized despite earlier attempts and often grand claims. We imagined a future model in which thirty students in the same program would master the same skills and knowledge, but have thirty different experiences, pathways, and content, each having learning experiences precisely tailored for them. We wondered if we might actually leapfrog personalized learning, that elusive goal of ed tech companies, and develop precision learning, the kind of n-of-1 understanding of a single learner that precision medicine was seeking with patients.

This was a profound notion. Education has largely been designed for the Industrial Age, with its need for large numbers of literate workers with higher-order skills. Needing to scale and measure performance, education adopted a one-size-fits-all delivery model that favored standardization and efficiency. If thirty students are enrolled in the same program, they will largely have the same experience, follow the same sequence of topics and activities, digest the same content—no matter what they know, how they learn, how they think and feel, regardless of their context and conditions. At almost every level, education is designed around the thing to be learned and how the instructor wants it learned (which is usually a reflection of how *they* learned it). George, one of the world's leaders in learning science, was convinced that AI would give us the tools to build powerful and dynamically updated learner profiles for each student, profiles that would serve up learning with a precision and efficacy that we have never seen in education. *Personalization*, the Holy Grail of education, would be in our grasp. Then we could move from one-size-fits-all education to student-centered precision learning, a paradigm shift. We could change the world of education.

Turns out we were thinking too narrowly. Tanya, a clinical psychologist with a thriving practice, was listening in and absorbing it all, then sent a kind of manifesto or white paper that just floored me. It was titled *Connectivism.AI: The Art of Being Human in the Age of Machines*, and she had written it when thinking about the ways AI might be used to address the mental health crisis sweeping the United States and much of the world, especially among young people. As her paper, written with urgency and a mix of despair and hope, explains:

> *Education/learning/schools were developed to meet different challenges and needs of a period long past. If we look at schools as what prepares people best for the future, where most of the tasks of "knowing and knowledge" are done by the brain/smartphone in our pocket, then the entire need for an education is different. We are holding on to this old model past when it is relevant, like keeping sets of encyclopedias way past when they were useful. We got arrogant in our cognitive superiority ignoring all other ways of knowing and being that exist in the more than human world around us. But AI usurps our cognitive superiority. Humans will need to find our place in a world where being the cognitively superior agent is no longer our role.*
>
>   i. *People have been circling around this awareness for a while.*
>
>   ii. *People are looking to reconnect to other ways of being. More connection to self, more connection to nature, more connection to others.*
>
> iii. *Our isolated, piecemeal mastery and domination of the world is a failure.*
>
> *By ignoring the larger systems at play we have begun to destroy them.*
>
>   1. *The environmental system*
>
>   2. *Our mental health*
>
>   3. *Our food system*
>
>   4. *Our body system*
>
>   5. *Our communal systems*
>
> *AI may be a huge opportunity to rebalance these systems as it will free up time to teach and learn and interact differently—there will be more time for being human in a complex system, if AI is handling the mundane details, drafting our documents, doing our computation, etc.*

I knew George well but had never met Tanya. I called her and we talked through her ideas, sharing our sense that the world is increasingly broken, that we have to use technology not to replace human engagement and relationship, but to enhance it. Her vision took us from a concentration on education to a broader focus on human development. I knew she had to join us, and our duo became a trio.

I shared our thinking with the SNHU Board of Trustees and argued that AI would change the educational landscape and could lead to the next generation of SNHU's learning model. The board eventually approved a $25 million investment in what became Matter and Space, the AI and education company George, Tanya, and I cofounded in June 2023. We started the company during my final year at SNHU. Two years into the work, on the eve of our September 2025 pilot launch with one thousand students across eight universities, SNHU decided to bring the platform in-house for its sole use, seeing it as a next-generation platform for its massive student population. George, Tanya and a small internal team continue to work on the platform at the time of this writing.

While we started out to build a learning platform, we came to call it a *human development* platform. At the heart of the platform, which we called Learning Environment 1 or LE-1 (Ellie to our eventual student testers), is a learner profile George originally sketched out on a napkin and that ingests myriad data, updated in real time, shaping each student's experience in the moment. LE-1 integrates learning, well-being supports, and soft-skills development, and does so employing multiple LLMs, agents, knowledge graphs, and machine learning along with generative AI. It integrates with and pulls data from wearables, such as the Apple Watch, Fitbits, and WHOOP bracelets, and we were working on integrating smart glasses, which we believe will be the key device of the future, linking users and their AI to the physical world around them. We were building LE-1 to be radically affordable, given our commitment to serving those who cannot afford postsecondary education and addressing the global need for affordable, high-quality education. LE-1 was our attempt to reimagine education as a dynamic system that develops the whole person, with broader human impact than the mere learning of content.

That broader vision for how AI might transform the world by democratizing high-quality education in ways that could make students feel seen and well supported was rooted in our own experiences. As immigrants and first-generation college graduates, George and I have experienced firsthand the power of education to transform a life. I can name the three teachers who had the most impact on me—one in sixth grade, another in high school, and one in college—and it was access to affordable, high-quality postsecondary education that opened career and other opportunities for me. I also know my experience is increasingly out of reach for too many people in the United States and globally. Too many learners do not have a teacher who makes them feel like they matter, who mentors them the way I was mentored. Too many people cannot afford high-quality educational programs and, in much of the world, there just are not enough seats. Moreover, higher education remains largely built for traditional age students pursuing two- and four-year degrees.

I had worked hard to address this shortcoming of the postsecondary system during my time leading SNHU. It grew to be the largest university in the United States because we focused on learners juggling full-time jobs, parenting, and too little money. They chipped away at degrees over time, like the long-haul truck drivers who would park their rig every night at a rest stop, then open their laptop, log on, and be a student for a couple of hours before getting some sleep. They joked that they earned their degree "in forty-eight states." There were also the refugees we educated in camps in Rwanda and Malawi, living on food rations and dreaming of life outside the camps. I realized we were not only in the business of education but also in the business of hope. Matter and Space came about because we believed we could use AI to give any learner access to the best knowledge and instruction, to the support for their well-being, to the soft skills necessary for success and advancement, and do so at a fraction of the cost of most education. And that we could harness personalization and AI's dialogic nature to make students feel like they matter when little else around them sends that message.

But this book is not about the two years we spent building Matter and Space and its breakthrough LE-1 platform. We write to share what we are learning about AI and, more important, to pose the big questions with which we are grappling. While our personal experiences and desire to make sure the transformational educational experiences we had are available to all, the questions we take up are situated in tectonic historical and societal shifts. Our larger contextual framing comes from Carlota Perez, the British Venezuelan economist and author of the seminal *Technological Revolutions and Financial Capital: The Dynamics of Bubbles and Golden Ages*. She reminds us that we remain in the turbulent transition period of a revolutionary, world-changing technology in which the pace and nature of technological change is outpacing our ability to evolve our thinking and systems. She would say we have been here before: a dazzling new technology bursts onto the scene, attracting torrents of money, lionizing its inventors, and promising to change everything.[6] And it does, but not for everyone, and not all at once. The early days, the "installation period" in her framework, are a heady mix of speculation, experimentation, and disruption. Financial capital runs the show. Fortunes are made. But the institutions, policies, and cultural norms built for the last era lag, ill-suited to manage the upheaval.

That is where we are with AI. The large AI companies are releasing new versions almost monthly, handling ever more complex tasks and moving beyond text to video, audio, vision, robotics, and instant translation while building massive data centers costing billions, eating up energy, and damaging the environment. The technology is racing ahead—generating code, analyzing medical images, tutoring students, and outperforming humans in myriad domains—while our rules for using it responsibly are being written in real time (and not very well). Witness the campus debates about banning AI in courses and its implications for academic integrity. Our pedagogy and assessment practices have not kept up. In the meantime, certain jobs are already disappearing or changing beyond recognition. Skilled professionals are wondering whether the thing they have spent a career mastering will still matter. A senior executive at a global bank shared that the company plans to cut its workforce by 30 percent over twenty-four months. The unemployment rate of recent computer science graduates is twice that of art history majors.[7] Meanwhile, dizzying wealth and influence are concentrated in a small circle of companies and investors.

Perez warns that this is the dangerous stretch. The benefits of the new technology are real but unevenly distributed. Disruption hits hardest where skills are slow to adapt and safety nets are thin. Social tensions grow. In past technological revolutions, this is when political backlash, mistrust, and even outright resistance to the new paradigm surged. These are periods when wealth inequality becomes extreme, populism rises, wars erupt, strongmen leaders emerge, monopolies dominate, and voter discontent increases. In every democratic election around the world last year, the incumbent party was deposed, whether left or right, except for Mexico and South Korea, and South Korea changed over months later after a political crisis. Sound familiar?

We are in the AI earthquake and it will not be soon over. The hopeful turn, in Perez's telling, comes only when we cross the bridge into the "deployment period," when the technology's potential is matched by the policies, education systems, and institutional reforms that let the benefits flow more broadly. In the Industrial Revolution, it was public schooling, labor protections, and infrastructure investment. In the Information Age, it was the spread of personal computing, the internet, and new industries that created millions of jobs.

For AI, the same challenge looms: building the guardrails, training systems, ethical frameworks, and equitable access that turn a powerful, sometimes disorienting invention into the backbone of a shared prosperity. Until we do, Perez would say the future will feel less like an open door and more like a locked one, its key held by a few.[8] But with AI, there is a twist at which Perez's earlier case studies can only hint. Steam, electricity, mass production, computing—each expanded what humans could do. But none directly stepped into the territory we have long claimed as ours alone: the capacity to reason, to imagine, to decide, to connect. AI is not just a faster loom or a bigger engine. It is an encroachment on intelligence itself, the thing we have used for centuries to define what it means to be human.

That is why AI feels different. It is not just an economic disruption; it is an existential one. If a machine can write a persuasive essay, diagnose an illness, or even feel like a friend or therapist, we start to ask uncomfortable questions: What is left for us? What does human judgment mean in a world where algorithmic decisions outperform our own? Where do we draw the moral lines about what only a person should do? How are we indeed different . . . and is that enough?

Perez would optimistically point to the eventual turning point ahead, the moment when we move into the deployment period and the technology's benefits are finally unlocked for the many, not just the few.[9] In every previous technological revolution, the chaos of the transition gave way to a new Golden Age, in Perez's telling. That critical transition required new laws, public education, infrastructure, and institutions designed for the new reality. With AI, it will also mean building ethical frameworks, guardrails, and cultural norms that decide not just *how* the technology is used, but *what parts of being human we refuse to outsource*. Our hope is that education—high-quality, personalized, engaging, and relevant—is available to all and treated as a human right, part of a possible new world order characterized by abundance and not the scarcity that today limits access, a greater range of options, and affordable choices for far too many learners.

Until then, we are in the messy middle, where the possibilities are thrilling, the risks are disquieting, and the future feels less like an open invitation and more like a debate over what it even means to be invited at all (and many would like to be left off the guest list). I had the opportunity to have dinner with one of the world's leading and best-known technologists (Chatham House rules prohibit sharing the name) and he said, "If I had a switch that could pause AI development to let us catch up in our thinking, I'd use it right now. We do not know how to rethink society for the world of abundance that AI will create." The man is a techno-optimist: he believes AI can create prosperity for all, eliminate the need to work, and address intractable challenges such as climate change, curing diseases, and extended life spans. In that sense, even with deep concern over bioterrorism and AI warfare, he echoes other techno-optimists like Ray Kurzweil, Sam Altman, and Marc Andreessen, author of the much-debated *Techno-Optimist Manifesto*, but he is more thoughtful about the existential questions of how we will derive meaning, how societies then operate, how economics is redefined. "How much baseball can we watch?" he quipped.

Those are the big questions, the province of philosophers, psychologists, sociologists, ethicists, and theologians. In a keynote talk I gave on AI and education in Cuzco, Peru, in 2024, I turned to the Catholic priests in the audience, of which there were many given the large number of Catholic universities across Latin America, and said, "Fathers, this is your time. You have always been about the big questions like 'Why are we here?' and 'What is our purpose?'" And while I was admittedly pandering a bit, I was also sincere and they nodded in happy affirmation.

The context and questions are a reminder that we are in the middle of a complex societal transition, full of unknowns, and that the technology fueling this change is evolving at warp speed. What we can say of the technology today is likely to be outdated tomorrow, but we can ask good questions. Indeed, certainty at this moment should be suspect and we simply need better questions, or the right questions. In the chapters that follow, we share the questions with which we grapple as we try to *rethink* education and human development for the new world order, as Antonio Gramsci would call it.[10] Because education should be about the important questions of purpose or vocation (and by extension work and career), community and society and one's role in them, and the tools for a life of flourishing, meaning, and impact, we will inevitably link back to the larger context just outlined, but through the lens of one key question: *What is the university that the world needs next?*

The chapters that follow address the hardest questions with which we wrestle as we built the Matter and Space platform:

- What will be the future of work and how will that then reshape what universities offer?

- Will education move from an epistemological to ontological focus, worrying less about what learners need to know and more about how learners need *to be* in the world?

- What does it mean to live alongside a new form of alien being, one that is more intelligent and seems human in many ways, and in a world in which many will prefer their AI companions to their human companions?

- Can we make high-quality, personalized education available worldwide for free or almost free?

- What of the incumbent system? How does it evolve? Will we need it or want it in the future? What will be the role of the faculty member in the future?

Are we techno-optimists? It depends on the day. As educators, we think these are the most fascinating questions we can explore. As people whose lives have been transformed by education and amazing mentors and teachers, we love our industry, but we also are painfully aware of how it is falling short. It leaves too many people behind, remains too far out of reach of too many globally, lies increasingly out of step with what society needs, and is often overly occupied with itself and not focused enough on students. Rethinking education for the emerging world is the most powerful tool we have for getting the Age of AI right.

We could sit, wait, and wring our hands, waiting to see how this all plays out. After all, we already know all the ways the world could get it wrong. We have seen *2001: A Space Odyssey*, *Her*, and *The Terminator*. Or we can try and shape the future, at least our corner of it, and be proactive. In the pages that follow, we invite the reader to join us on the journey, think about the questions, and imagine with us a future in which education and AI can help fix a world that feels deeply broken.

Some final notes to the reader. While the three of us collaborated on the book in its entirety, I (Paul here) have taken the role as lead author, thus the first-person point of view to simplify matters. George contributed much of the chapter two primer on AI, and most of the thinking for chapter five on precision learning, while Tanya is the intellectual force behind chapter four. I wrote the remaining chapters, and George and Tanya's comments and edits greatly improved them. That said, any shortcomings of the book are undoubtedly my fault; George and Tanya are the best of collaborators. We derived so much of what we share in this book from our actual *practice*, our working to build much of what we outline in the book. We have been immersed in AI these last two years (far longer for George) and write as practitioners, which means we often got things wrong along the way when working on LE-1. We are grateful for the missteps, as we often learned more from them than from our successes.

I used AI in the writing of this book, and because I believe that transparency in the use of AI is an ethical imperative, I identified those uses in the AI Disclosure.

Perhaps the most painful aspect of AI writing (at least to me) is the way it has ruined the em dash for everyone by overusing it. It has long been one of my favorite grammatical devices but has become a signal that screams "AI wrote this text." I assure you, if you come across an occasional em dash in the pages that follow it was not produced by AI. I reluctantly and mostly abandoned the em dash, but at times could not help myself.

The chapters that follow can be read as a series of essays, but we hope the through lines are readily apparent. The chapters are as follows:

- Chapter two: A primer on AI. I debated including this, but I see so much careless throwing around of technical terms and a frequent casting of AI as a *new* technology, I thought it was important for readers to have a brief overview of what is, in fact, a seventy-year-old technology (a combination of technologies, really) that did not start with the release of Chat GPT 3.5 in 2022.

- Chapter three: This chapter outlines the idea of an emerging "care economy" as white-collar jobs in the information economy get displaced by AI, and the need for higher education to more fully embrace its mission concerning ontological and existential questions, a mission more about developing better humans than training for jobs.

- Chapter four: We explore anthropomorphism, mattering, and what it means to live and learn alongside this new alien intelligence we have created.

- Chapter five: I outline the ways AI can enable us to leapfrog mere personalized learning to breakthrough precision learning, fundamentally changing the longstanding one-size-fits-all models of education that emerged with the Industrial Age.

- Chapter six: I get more focused on the university itself and the ways universities need to address the emerging Age of AI for which they are preparing their graduates.

- Chapter seven: Here the focus is on the role of the faculty member, especially as knowledge transfer is more effectively delivered by AI-driven platforms like the one we were building at Matter and Space. I posit the idea of the faculty member as a curator of community, more focused on relationship and human development than a discipline.

- Chapter eight: This one moves from universities to the larger ecosystem in which they exist, including questions of credentials, assessment, accreditation, and equity and access.

- Chapter nine: I finish by opening the aperture again, challenging our sector to rethink its role, to broaden the definition of education, and to elevate itself, to dream bigger, about what universities can be in a new world order.

George, Tanya, and I are excited about the possibilities for reinventing higher education, realistic about just how darn hard that will be, sober about the pains of the transition period, and absolutely convinced that a higher education system that moves from workforce training to developing flourishing human beings is the answer to an increasingly broken system of learning in an increasingly broken world. Tectonic shifts are underway, the ground is shaking beneath our feet, and while we fret about the damage being done, we are inspired by what is possible.

A Brief History of AI for the Layperson

The arrival of ChatGPT 3.5 in November 2022 felt less like a technological upgrade than a global lightning-strike event, and not long after student essays were uncannily improving, trustees were asking university presidents about their artificial intelligence (AI) strategies, and faculty were debating its use in their courses. It seemed like AI happened almost overnight. Indeed, technological revolutions often have a catalytic moment like that one, a point in time when multiple innovations come together in some new "change-the-world" combination. The release of Chat GPT 3.5 to the public likely will be remembered the same way historians characterize the 1829 test of The Rocket, the breakthrough steam locomotive that started the railway age and a transportation revolution. It too felt as though it appeared overnight, but in fact it united multiple decades-long technological advances in precision metalwork and machinery, boiler construction, wheel design, and more. To most of us, ChatGPT's arrival felt like waking up to hear an alien had landed on the planet. We did not see that one coming.

But in truth, AI has been around for decades. What we think of as AI is the combination of multiple innovations developed over time. That journey has not been a steady march of progress; it is a rocky, volatile tale dating to the 1950s and characterized by dizzying highs, catastrophic crashes, and breakthroughs fueled by brilliant ambition, brute-force computing, market demand, and, increasingly, geopolitical necessity. AI is becoming a "foundational technology" that becomes part of the texture of day-to-day life, like electricity, and is changing everything. Like those breakthroughs that eventually created The Rocket, ChatGPT and the other generative AI models transforming our world are the result of multiple technological advances over decades, sparked by one provocative question.

The Foundation Era (1950–1974): The Philosophical Spark

This period began with Alan Turing's fundamental question, "Can machines think?" Turing was the brilliant British mathematician and computer scientist (made even more famous by the 2014 movie *The Imitation Game*). Rather than trying to define "thinking," he suggested a practical test, later called the Turing Test: a human judge engages in a text-based conversation with two unseen entities, one human and one machine. If the judge cannot reliably distinguish which is which based on their responses, then the machine could be said to exhibit intelligence, at least in a behavioral sense.[1]

At the landmark 1956 Dartmouth Conference, researchers christened the field and set a wildly ambitious agenda based on the "conjecture that every aspect of learning or any other feature of intelligence can in principle be so precisely described that a machine can be made to simulate it."[2] Participants rejected the term *automata studies* in favor of *artificial intelligence*, the name preferred by John McCarthy, the pioneering "godfather of AI" who later founded the Stanford AI Lab. It reflected his ambition to build systems that could think and reason like humans, not merely automate simple processes. Early work rooted in symbolic AI by McCarthy, Marvin Minsky, Claude Shannon, and Nathaniel Rochester resulted in the birth of early neural networks like the wonderfully named Perceptron, one of the first attempts to teach a machine to "learn" from data the way a brain does. It was invented in the late 1950s by psychologist Frank Rosenblatt, who imagined it as a computational model inspired by biological neurons. At its core, a Perceptron takes inputs (like pixels in an image), assigns weights to them, adds them up, then applies a simple rule to decide on an output (for example, "this is a triangle" or "this is not a triangle"). By adjusting the weights during training, the Perceptron could learn simple pattern-recognition tasks. In 1958, Rosenblatt demonstrated it on specialized hardware called the Mark I Perceptron, which caused a wave of excitement; newspapers proclaimed that machines that could walk, talk, and reproduce were on the horizon.

But Perceptrons had major limitations. In 1969, Minsky and Seymour Papert showed that single-layer Perceptrons could not solve even basic logic problems like distinguishing when *only one* of two inputs is true. That critique stalled neural-network research for years. Later, multilayer versions and the backpropagation algorithm revived the field in the 1980s, ultimately leading to today's deep learning systems. In that sense, the humble Perceptron is the ancestor of modern AI, from image recognition to large language models (LLMs).

The Institute for the Future in Palo Alto, California, uses the term *artifact from the future* to describe things in our world that function as early prototypes or are launched, then fail as commercial products, but foretell a future breakthrough.[3] Apple released its Newton MessagePad handheld device in 1993 with great fanfare. It was a dismal failure, selling about two hundred thousand units over almost five years until it was discontinued, but it anticipated the PalmPilot, the BlackBerry, and eventually the iPhone. It was an artifact from the future. In that respect, we can view the Perceptron and other early attempts at thinking machines as artifacts from the future, now our present—not only in setting a technological foundation for modern AI but also in the politics and global competition to control AI. Early efforts occurred during the Cold War, and the US military saw in AI huge potential for military applications, the reason why the US Navy funded Rosenblatt's research. The government's support of the work anticipated the modern arms race between the United States and China, each competing for AI superiority.

AI Winters and Commercial Spring (1974–1995): The Crash

The inevitable happened: overblown promises met technical reality. Machines did not learn language with human nuance, robots did not glide across factory floors with graceful autonomy, and computers stubbornly refused to replicate the pliable reasoning of even an average person. Politicians and funders, once dazzled by the confident rhetoric of early AI pioneers, began to see diminishing returns and rising skepticism. By the late 1970s and again in the late 1980s, the money dried up. What followed is remembered as the first major AI winter, a period of disillusionment when research budgets evaporated, headlines turned cynical, and the field's once-lofty ambitions retreated behind the closed doors of shrinking labs and shuttered projects. This contraction, painful as it was, was a clarifying moment. Researchers had to narrow their focus from sweeping visions of artificial general intelligence to something far more modest and grounded: systems that could perform specific bounded tasks exceptionally well. Rather than trying to build machines that could think broadly like humans, they built systems that could diagnose diseases, schedule airline crew rotations, and recommend investment strategies. They became the first commercially viable applications of AI, translating academic theory into the tools industry could use and trust.

It was AI's first big lesson in humility, a recognition that imitation of human thought is a long, iterative climb that must begin with the mundane before it reaches the miraculous. Although these expert systems eventually ran into their own limits—they were brittle, inflexible, unable impossible to adapt or generalize—their success proved that AI did not have to rival the human mind to matter, but simply solve real problems better, faster, or cheaper than humans alone. That pivot laid the groundwork for an AI renaissance decades later, when computing power, data availability, and new learning techniques finally caught up to the field's early imagination.

In this phase, focus shifted to creating expert systems, which function like a hyper-specialized digital consultant. Instead of trying to build a robot that could invent a new kind of motor, researchers built a program that could diagnose problems in existing motors based purely on a rigid set of rules provided by a master mechanic: *IF the engine is hot AND the fan is not running, THEN check the thermostat.* In education, an early expert system might have advised a student on selecting a major based on an inflexible set of rules about grades and course history. I tried my hand at such an expert system as a graduate student in the 1980s. When I was completing my doctorate at the University of Massachusetts in Amherst, I was in the English Department studying the impact of digital technology on literacy. Beverly Woolf, a professor of computer science and a member of my dissertation committee, was doing early work using intelligent tutoring systems to train nurses. Despite some head scratching by the English Department chair, I persuaded him to let me learn Turbo Pascal as one of my foreign languages (one needed competence in two foreign languages at the time, and Turbo Pascal was indeed *foreign* to my tweedy department chair). The early object-oriented programming language became popular in education and enabled me to play with prototypes and create step-by-step "coaching" for activities like revising an essay. These early systems were impressive for their time, but only within single, narrow domains, and they could not handle new information or nuances.

As before, the history of AI is always interwoven with grand aspirations for commerce, research, and warfare. A crucial geopolitical moment occurred in 1982, when Japan's Fifth Generation Computer Systems project sparked a major reignition of state-level investment globally. This was an early and clear sign of AI's perceived threat to the international order and signaled the end of AI winter. Research labs were sowing the first buds of a new spring: a quiet academic revolution, known as connectionism, and developing new algorithms, such as backpropagation, that created the foundation for neural networks and the AI products now in the hands of consumers.

The Foundations of the Modern Era (1995–2011): The Triple-Curve Convergence

In 1993, I took a leave of absence from my faculty role at Springfield College to lead a new education technology unit at Houghton Mifflin Co., a storied Boston-based publisher that had published Nathaniel Hawthorne, F. Scott Fitzgerald, and Robert Frost. Publishers were rushing to stand up "new media" units to take advantage of multimedia platforms that linked various kinds of information. I was leading a cutting-edge new media lab and classroom at Springfield, and Houghton Mifflin editors spent two days observing my students working in new ways, such as creating multimedia "papers" on William Shakespeare's plays, linking in nonlinear ways not only text but video clips, animations, and more. The editors asked me to help them think about the future of their industry, which was an invitation impossible to turn down and that became a three-year stint at a time of exhilarating technological change. I remember coming into a Houghton Mifflin leadership team meeting and telling everyone that I had just used for the first time a browser named Mosaic, which provided access to disparate online information on a thing called the World Wide Web. I also was obsessed with a new online computer game called Myst, which made the old Atari and Nintendo games we all knew seem primitive. My colleagues met my enthusiasm with a bit of eye rolling.

What I did not recognize was that in the rapid growth of the web, the related collection of data, the development of sophisticated online games like Myst were underlying technologies that would later set the stage for modern AI. How so?

- **The explosion of data fueled by the public internet.** The rapid expansion of the internet led to an unprecedented increase in the amount of data generated and shared online. This massive influx of data provides the raw material necessary for training AI models, enabling them to learn and make predictions more effectively.

- **The maturation of powerful statistical machine learning algorithms.** Over the years, these algorithms have evolved to become more sophisticated and efficient, allowing for better data analysis and interpretation. This maturation has enabled AI systems to recognize patterns, make decisions, and improve their performance over time.

- **The harnessing of parallel computing via graphics processing units (GPUs), repurposed from the video game industry.** GPUs, originally designed for rendering graphics in video games, have proven to be highly effective for performing the complex calculations required in AI and machine learning. Their ability to handle many operations simultaneously has significantly accelerated the training of AI models, leading to advancements in the field. For example, the same year I joined Houghton Mifflin, Jensen Huang, Chris Malachowsky, and Curtis Priem formed the GPU-production company Nvidia in Silicon Valley (where else?) to serve the gaming industry. However, GPUs are ideal for AI because their thousands of parallel cores, high memory bandwidth, and supportive software frameworks make them far faster than central processing units, or CPUs, at performing the large-scale matrix computations that power neural networks, the foundation of AI. Those seeds have made Nvidia the most valuable company in the world—$5 trillion at the time of this writing, the first company to reach that number.[4]

We did not have to wait long for a preview of AI's coming cultural revolution. IBM's Deep Blue defeated world chess champion Garry Kasparov in 1997, making headlines the world over and showing that machines could dominate complex strategy.[5] Deep Blue was followed by Watson, which famously beat champions Ken Jennings and Brad Rutter on *Jeopardy!* Now our grandparents were talking about AI. Watson brought together innovations in natural language processing, enabling users to engage with everyday English in a way we now take for granted. It harnessed machine learning, massive parallel processing, and training on large knowledge bases. IBM saw healthcare as the first great commercial opportunity for Watson. But Watson Health struggled, since medical knowledge is messy, context-dependent, and often ambiguously documented. Those real-world challenges made translating natural language into actionable, reliable clinical recommendations far more complex than IBM anticipated, leading to errors, overpromises, and limited real-world adoption. While Watson Health never lived up to its promise (not the first or last time a technology company engaged in premature hype), Watson in many ways set the blueprint for much of what was to come. This period quietly created the necessary preconditions for the deep learning revolution to follow.

The Deep Learning Revolution (2012–2020): The Big Bang

Just as the Turing Test galvanized early research into AI, the lesser-known ImageNet Competition catalyzed the Big Bang of modern AI. The annual computer vision contest that ran from

2010 to 2017 challenged teams to build algorithms capable of accurately identifying and classifying millions of labeled images across thousands of categories. It became a pivotal benchmark for AI, especially deep learning: in 2012 a convolutional neural network, AlexNet, dramatically outperformed previous methods, demonstrating the power of neural networks and GPUs for image recognition, heralding the arrival of the modern era of AI. AlexNet's stunning victory provided irrefutable proof of the power of deep neural networks, unleashing an unprecedented wave of private investment and corporate research and development. Deep neural networks learn from examples, much like the human brain learns from experience. The networks are made up of many small processing units, called *neurons*, that connect in layers. Each neuron looks for tiny patterns in data—shapes, colors, sounds—and passes the information along so deeper layers can recognize more complex features. Rather than following a strict set of rules, a deep neural network adjusts its internal connections when it makes mistakes, a process known as training. Over time, it learns to make accurate predictions or recognize patterns on its own. Deep neural networks power many of the technologies we use every day—from facial recognition and voice assistants to language models like ChatGPT and recommendation systems on Netflix. In short, they enable computers to learn by example rather than by explicit instruction. Whereas my early and mostly failed attempt at an expert tutoring system for revising student writing required me to set rigid rules and procedural steps, a neural network learns by looking at examples, providing guidance, being told where it did well and it where it did not (the training portion of the work), then trying again and again and again until it gets very good at guiding student revisions. We now have machines that can "learn."

The impact of AlexNet, which revolutionized image classification, immediately translated to educational tools. Consider modern automated grading of assessments. Before 2012, it was impossible for a computer to grade a drawing or a hand-drawn diagram. Post-AlexNet, it was possible to train AI on massive sets of labeled images (e.g., student diagrams) and learn to instantly recognize and assess the correctness of a complex image, opening the door for new types of scalable visual assignments.

At the same time, the geopolitical backstory for AI is never far removed. The revolution accelerated in 2016, when DeepMind's AlphaGo defeated Lee Sedol, world champion of the strategy game Go. This was a seismic moment for the Chinese in particular, and not just because of Go's enormous popularity in China. It is a game of immense complexity, intuition, and pattern recognition, qualities long thought to be uniquely human. AlphaGo's four-to-one victory demonstrated that AI could master tasks requiring not just brute computation but subtle strategic judgment, a signal to China that leadership in AI could translate directly into economic, military, and technological power. Beijing responded swiftly, launching within one year a national AI strategy that explicitly framed AI as a tool of statecraft, setting a 2030 goal to become the world leader and cementing AI as a focal point of the US-China strategic competition.

In 2017, AI's crucial technical breakthrough arrived with the invention of transformer architecture. Transformer architecture solved the efficiency and memory problems inherent in prior neural networks, which used sequential models. Simply, a sequential model had to finish processing the second word in a sentence, for example, before it could process the third word, and so on. They were slow and prone to error (often forgetting the first words of a text when they finally got to the final words). In 2017, researchers from Google Brain and Google Research introduced AI transformer technology.[6] The core innovation of transformer architecture was the self-attention mechanism, which enables the model to process *all words* in an input simultaneously, rather than one by one, thereby enabling massive parallelism and accelerating training times. Self-attention calculates the relationship (or "attention score") between every word and every other word in the sequence, enabling the model to instantly grasp long-range dependencies—for instance, correctly determining what a pronoun refers to many words earlier in a sentence. No more word-by-word sequencing. With other component parts of the architecture, an encoder (for processing and understanding the input context), a decoder (for generating the output sequence), and a critical cross-attention layer bridging the two, the system could selectively focus on the most relevant parts of the input. Much in the way we as humans read, but with an ability to now digest massive amounts of texts. As in almost everything printed and digitized. Transformer architecture provided the scalable model underpinning for everything that followed, culminating in the 2020 release of OpenAI's GPT-3, which began the democratization of large-scale AI.

The Generative Era (2020–Present): Public Awakening

No event affected public awareness more than the launch of ChatGPT 3.5. Deep Blue, Watson, and AlphaGo had their moments in the sun, and applications like GPS systems and customer service chatbots enabled AI to seep into everyday life in ways not widely recognized or appreciated. But like The Rocket, ChatGPT 3.5 seized the general public's imagination, and it immediately found its way into day-to-day use. ChatGPT 3.5 signaled a new era of AI, defined by the rapid development of multimodal models (handling text, images, and video) and the emergence of autonomous AI agents that are beginning to reshape labor markets and creative industries.

The era's tools move beyond simple automation. An AI agent is more than a chatbot; it is a piece of software that can take a high-level goal and break it down into steps, execute them, and learn from feedback. AI can be used for research, for editorial work, for creative work, for customer service, for processing vast amounts of information. In education, this means going from a program that just corrects grammar (old AI) to a generative AI that can draft an entire personalized syllabus for a student, create five new assignment prompts with varying difficulty levels, then generate a reading list from a massive online library—all based on a single prompt from a faculty member. It can role-play with a student, offer feedback, provide around-the-clock, never judgmental tutoring on any topic, and walk a prospective student through applying for admission and navigating the complexities of financial aid.

The launch of ChatGPT 3.5 was AI's "rocket" moment, assembling previous advances in a startling and powerful new way that is changing the world. As Carlota Perez points out, AI is the latest chapter in what is really the digital technology revolution that had already reinvented the world before November 2022.[7] Yet it may be the chapter that transforms the world and human life in more profound ways, because it is the first technology that introduces to the planet an intelligence greater than ours, at least in many realms of intelligence, and one whose black box thinking is not clear or apparent to us. Those of us old enough to remember duck-and-cover drills at school during the Cold War know AI is not the first technology to threaten our survival as a species. But that threat of annihilation came from other humans (mostly the Russians in those simpler times). Today's AI doomsayers point out that the enemy we perhaps face is vastly smarter than us in many ways, and we may not control its "thinking." It begs for regulation and oversight, for global agreements on what is in and out of bounds.

The global race for AI dominance is between the United States and China, and there is no hint of any global pacts and treaties that might establish guardrails on development and deployment. At the time of this writing, the Trump administration has pushed back against any regulation that might slow AI's development for fear of falling behind China, which has the resources and technology to keep up with America's AI innovations. Big AI companies like OpenAI, Google, Anthropic, Amazon, and others are building massive data centers, driving up demand for electricity, using up massive amounts of water for cooling, and extracting an environmental toll on communities that can hardly withstand the impacts. A single large data center can use the same amount of power needed by a good-sized American city, and the billions required for their construction are inflating the United States' gross national product. The value of the companies driving AI innovation, from chip manufacturer Nvidia to Open AI to Alphabet, is measured in the trillions of dollars. Economists fear an AI economic bubble, an overheated valuation of companies that are still losing massive amounts of money on their AI bets. Companies are still trying to translate the power of AI into new business models and productivity and profit gains that match the hype. Despite its power to do dazzling work, such as identifying in just days the protein structures that teams of medical researchers spent years considering, AI can still provide mind-boggling "hallucinations" to seemingly simple requests, mistakes that make us doubt its intelligence (like many people we know). Large AI companies are fighting for world domination—no hyperbole here—and thus make announcement after announcement, some of them substantial, some merely meh, and others embarrassing. Even when their advances yield sycophantic agents or depictions of slaveholding founding fathers as Black, their fierce competition is driving rapid technological advances, all while driving down the price of computing.

There is much debate about the limitations of current AI models—whether LLMs are capable of overcoming their current limitations and getting us to artificial general intelligence (AGI), and if so-called large world models (LWM) are the next chapter in the story of AI. Some of the best minds in the field are betting on LWMs, including Yann LeCun, former chief AI scientist at Meta; Ilya Sutskever, the former chief scientist at OpenAI; and the previously mentioned Fei-Fei Li, the creator of ImageNet. LeCun has been most vocal about the limitations of LLMs, which are built on massive volumes of text data, but does not possess or generate any model of how the word works. LWMs build models of how the world works, observing and simulating and predicting, much as humans do. As LeCun points out, "The largest LLM is trained on about 30 trillion words, roughly 10^{14} bytes of text. That seems enormous, but a 4-year-old child who has been awake about 16,000 hours has also absorbed roughly 10^{14} bytes—just through their eyes. . . . One learns physics. The other learns autocomplete."[8] The new companies created by world class researchers like LeCun, Sutskever, and Li are attracting billions of dollars in investment and may eventually deliver the next powerful iteration of AI, the one that comes closer to the way our brains actually work.

In the meantime, AI as we know it today is ushering in a new world order. It raises existential questions about what it means to be human, to think, and to work. It is already reinventing warfare, commerce, scientific research, medicine, and more. The focus of this book is education, particularly postsecondary education, and in the chapters that follow we explore the possibilities, grapple with the challenges and vexing questions, and start to map out a new territory whose contours are still unfolding. I try to avoid the technology and focus on the bigger questions. Whatever one says about the technology today will likely be out-of-date by the time this book is in readers' hands, like those early maps of the new world that showed mermaid and sea monsters and vast expanses of empty lands. But the big questions about how AI is changing society and the workforce, and thus education and the university, are taking shape as we grapple with what it means if machines really can think—the question that started it all.

The Idea of the University Revisited:
From Epistemology to Ontology

When my daughter Emma was about eight years old, she was already a veteran of the many campus receptions and dinners we hosted at the president's house at Marlboro College. Weary of the social niceties required, she said she wanted a name tag revealing answers to the four questions she was asked repeatedly by kind strangers: "Emma. Third grade. Yes, I like school very much. No, I don't know what I want to be when I grow up." As most high schoolers know, that last question morphs into "What do you want to study in college?" Students feel pressure from a young age to have an answer. Indeed, surveys of college-going students are conclusive: the number one motivation for attending college is to land a good job and ensure a rewarding career. Asking "What will be your major" is just another version of "What do you want to be when you grow up?" only with more urgency for the college bound. We cannot talk about the future of postsecondary education without talking about the future of the workforce.

While society asks higher education to perform many jobs—conducting research and scholarship, providing sports entertainment, acting as an economic engine in local and regional economies—preparing people for the needs of the workforce while opening economic opportunities to graduates is job number one. In the knowledge or digital economy that dominates the United States and most developed economies, universities have become knowledge factories preparing knowledge workers for the knowledge economy. Unfortunately, universities spend far less time on actual skill development, on ensuring that students know what *to do* with what they now know. As work has become more specialized, so too have academic disciplines (squeezing out general education wherever possible). In many ways, today's graduates know *more* about their chosen discipline and *less* about the world.

When students finally decide what they want to do for work, when they can answer the question of *what they want to be when they grow up*, the associated college major effectively responds, "Then this is what you need to *know*." As such, universities are in the epistemology business, consumed with questions of knowing and content and student work that signals knowing. Faculty, academic departments, and curriculum committees endlessly debate course content, program content, pedagogy, assessments, and the latest thinking in the disciplines—all attempts to answer the question of what students should *know* and what is required in the major. The exceptions are majors in the humanities, which account for less than 10 percent of degrees conferred, perhaps because there is rarely a direct link to a good paying job.[1]

Ironically, even as universities have become far more focused on professional preparation over the last three decades, their graduates seem less ready for the workplace. Employers routinely complain about recent graduates being unprepared for the work expected of them. The chief human resources officer at one of the country's largest healthcare systems told me that the company used to place newly graduated nurses on the hospital floor in just days. Now the system hosts an internal "academy" that provides new nurses with an additional six months of training before they are deemed ready to take care of patients. "Nurses are coming to us without the skills required in modern nursing," he complained. While some of what he cited had to do with operating modes at his hospitals and clinics, he became most passionate when talking about new nurses' interpersonal skills, emotional intelligence, and problem-solving. There are many reasons for this growing skills gap:

- Employers' demand for day-one-ready employees and their reluctance to fund training programs for new hires

- The speed of change across many job categories, a rate of change that university academic departments cannot match

- The longstanding inadequacy of the employer-higher education dialogue and collaboration
- The plummeting level of preparation to do college-level work among high school graduates, reflecting some combination of diminished K–12 standards, grade inflation, and a global and alarming decline in academic performance tied to digital distraction and the pandemic[2]

Higher education has sought to address the problem by offering more internships, adopting competency-based education models that focus on demonstrable skills, and creating "authentic" assessments that mirror the expectations of the workplace. Take two fields in which performance might be life and death: healthcare and aviation. Education and training providers have long employed clinicals for the former, and for the latter many hours in simulators, then the right-hand seat in the cockpit before allowing a pilot to captain a flight. Both happen under the watchful eye of an experienced practitioner: a nurse supervisor or a senior pilot. This kind of experiential learning, rooted in actual practice, is lengthy and expensive and thus rare. Compounding the problem: the rise of digital and hybrid learning models has increased the time students spend working alone, reducing opportunities for sustained interpersonal interaction, collaboration, and mentorship. Coursework has become more modular and siloed, further fragmenting the kinds of shared problem-solving experiences that cultivate teamwork, communication, and adaptability.

These educational changes reflect broader social shifts: people today spend far more time alone than in past decades, and the proportion of waking hours devoted to solitary activities has been rising steadily since the early 2000s. The pandemic accelerated this trend, but the underlying pattern was already underway. As opportunities for in-person connection and collective work have declined, the social fabric and commensurate skills that once developed naturally through shared experiences has weakened.

In this context, universities are trying to cultivate collaboration, communication, and emotional intelligence within a culture that prizes productivity, digital engagement, and self-direction. The conditions in which students learn often work against the very interpersonal capacities they are meant to develop. As a result, the skills gap is widening as employers echo the frustration of that healthcare system chief human resource officer and report that many graduates lack the interpersonal strengths once taken for granted: the ability to work effectively with others, communicate clearly, take initiative, and solve problems collaboratively.

The Impact of Artificial Intelligence on the Workforce

Enter AI (artificial intelligence). Employers and universities alike are struggling to understand how to prepare students for a workforce undergoing profound change. It is expected that 60 to 70 percent of all jobs will be redefined, which means the undergraduate majors that prepare students for those jobs have to be redefined.[3] More dire predictions argue that whole job categories will be performed by AI, which can work twenty-four hours a day, will not complain or get distracted, and will never ask for a raise. Such predictions see AI doing to white-collar jobs what automation did to blue-collar jobs years ago. There is much debate on this question. Matt Sigelman of The Burning Glass Institute puts skills into two categories: (1) automation-exposed skills and (2) enhancement-exposed skills. As he points out, many jobs consist of both.[4] For example, the communications director for a small or medium-sized company may produce news releases, internal project updates, newsletters, and website content while also performing crisis management and strategic communications for the board and investors. The former lend themselves to AI automation, while the latter to AI enhancement. That communications person will have a redefined job more focused on work requiring judgment and wisdom. This kind of redefinition of jobs does not underestimate the substantial reshaping of the workforce by AI, but it is less alarming than more dire predictions.

Those abound. In a 2024 meeting on the topic, a former secretary of commerce commented, "I have a pit in my stomach when I think about how bad this will be. The AI shock to the workforce and to the economy will be ten times worse than was the China shock. And we already massively underestimate the level of unemployment and underemployment in the country." A senior leader from one of the big consulting firms said to me, "When I think about what our clients are telling us about their workforce plans, well . . . it terrifies me." Dario Amodei, the CEO of Anthropic, one of the main AI companies, has been willing to sound the alarm, and predicts that AI may eliminate 50 percent of all white-collar jobs over the next five years.[5] BGI hopes to create an AI tracking hub that can provide real-time insights into AI's impact on the workforce, as it is hard to get a clear picture and to separate job loss from AI versus economic performance or lower corporate spending and hiring due to volatility. What we do know is that organizations are racing to deploy AI.

Large AI companies are expanding capabilities and features at a breakneck pace, while employees are using AI as a productivity tool in myriad ways, and organizations are seeing a tidal wave of AI point solutions. A lot of employee use of AI is self-initiated as a personal productivity tool. Harvard's Project on Workforce has a generative AI adoption tracker and reports that 37.4 percent of employees use generative AI for their work as of December 2025.[6] Because we are early in the adoption process and employers are still bolting AI onto existing systems and processes, their reported productivity gains remain modest at 5 percent.[7] Job descriptions, a lagging indicator, are not reflecting a wholesale recasting of jobs through an AI lens. But talk to people doing the work, and you can feel the ground shifting beneath their feet. In marketing departments, for instance, content specialists who once spent days drafting copy now cocreate campaigns with AI tools that generate, test, and personalize messaging in real time—turning writers into editors, prompt engineers, and brand strategists. In software engineering, coders increasingly use AI copilots that autocomplete entire functions, accelerating development cycles and shifting the focus from typing code to designing system logic and reviewing machine-suggested fixes. Legal teams are quietly automating contract reviews and compliance checks, with junior associates spending less time redlining and more time interpreting the results AI surfaces.

A mid-sized architecture firm in Chicago recently used an AI-assisted design platform to generate and refine dozens of structural concepts for a downtown residential tower. What once took a team of six designers two weeks was compressed into a single afternoon.[8] Rather than eliminating jobs, the technology transformed them. Architects became curators and critics of AI-generated options, applying human judgment (there's that word again) to select and refine the most promising designs. Across industries, these shifts suggest that AI is no longer a distant disruptor but at least an immediate collaborator, one that is subtly rewriting workflows, redefining expertise, and accelerating the tempo of change inside organizations faster than job descriptions or productivity metrics can yet capture. I am in the camp that believes AI will do to white-collar jobs what automation did to blue collar jobs.

Though blue collar jobs will not be immune to AI either. Sigelman reports that one in five blue collar workers routinely use AI, while one in five of those holding only a high school degree regularly use AI at work. As AI is increasingly linked with low-cost, high-dexterity robots, some even see jobs in agriculture, construction, and manufacturing being further displaced. China's "dark factories" are an example of what is possible: companies use AI and robots so extensively in manufacturing that there are no humans on the factory floor, thus no need to ever turn on the lights. Serious people are asking if we are entering an age of AI-driven abundance in which humans do not have to work.

As AI takes over more of what is procedural, informational, and analytic, it throws into relief the dimensions of learning that remain distinctly human: creativity, connection, empathy, and meaning making. In this sense, AI is not only a catalyst for change but also a mirror, revealing where our systems of education have overvalued information and undervalued coherence. So many of our graduates are prepared to do routine cognitive work that is particularly vulnerable to AI automation. Amodei says that recent graduates and early career professionals, those in "administrative, managerial, and tech jobs for people under 30," are most at risk.[9] We are seeing entry-level jobs disappear and recent college graduates suffer higher unemployment rates than the general population, something we have not seen for decades. Parents of high schoolers, thinking about college and career choices, are frightened at the changes they see underway, worried about their children's futures. What should their kids choose as a major, and what work will humans do in the Age of AI?

Meanwhile, universities are struggling to give students and faculty even basic guidance and support for using AI, never mind a clear understanding of how their programs need to change. It will take some time to answer that question, as we remain in the early stages of the AI revolution. Then there is likely to be a counterrevolution, including a backlash and calls for greater controls. A senior executive of a global bank confided to me that internal plans call for a 30 percent reduction in its workforce over the next two years through the aggressive deployment of AI. If such massive job cuts become common, we will almost certainly see economic recession or worse, labor unrest, probably civil unrest. It might be argued that the 2023 Hollywood writers' strike was the first AI-related labor action. As people see their livelihoods increasingly under threat, we should expect calls for more regulation, curbs on AI deployment, and professional and licensing organizations fighting back. When doctors start losing their jobs to AI, do not expect the American Medical Association, the American Institute of Certified Public Accounting, the American Bar Association, and others to stand still.

There may also be other factors that slow the AI train barreling toward us. First, the technology is still developing, and the overheated claims about artificial general intelligence and super intelligence are giving way to more moderate predictions. Deep Mind CEO and Nobel Prize winner Demis Hassabis said in early 2025 that artificial general intelligence was at least five years away.[10] Existing large language models (LLMs) are astonishingly good at many things (and getting better), and yet surprisingly inaccurate at times in what we properly call *hallucinations*. As people find more uses for AI, we find more limitations. Some of them are complex, as seen when adolescents turn to AI when struggling with mental health issues. The media has reported on heartbreaking cases of teen deaths by suicide after flawed interactions with AI agents. Considerable technological questions about agent design, memory, context, and multimodal AI persist. AI, in the form of LLMs, has been trained on digitized texts, maybe most or all that exist, and is only now being trained on the physical world, what are called *large world models*, required for everything from robots building houses to improved autonomous vehicles and planes, to uses on the battlefield. We remain early in the AI revolution.

Second, even as AI technology evolves, or maybe *because* of how it evolves, we will inevitably see AI disasters. Some might be intentional, such as the prospect of bioterrorism enabled by AI and inexpensive desktop DNA synthesizers or biofabrication devices that can create or modify biological organisms from digital genetic code, as Mustafa Suleyman, the cofounder of Deep Mind and head of AI at Microsoft notes.[11] We have already seen the use of AI in financial crimes and political disinformation campaigns. In January 2024, voters in my home state of New Hampshire received AI-generated robocalls imitating President Joe Biden's voice ahead of the Democratic primary. It urged recipients not to vote, saying their participation would "only enable the Republicans." The voice clone was later traced to a political operative who had used an AI voice-generation tool, prompting the Federal Communications Commission to open an investigation and ban AI-generated voices in robocalls.[12] Some disasters might be unintentional, as AI systems with increasing control of infrastructure in energy or transportation, for example, endanger or costing human lives. All revolutionary technologies tend to fail and cause damage, necessitating regulation. When rail went from its origins in hauling coal up from mines to transporting humans across distances at never-before-seen speeds, fatal railway disasters were not uncommon until the standardization of time zones, rail gauges, and scheduling. Social media has taken a terrible toll on children (and on politics and society in general) and only now are we seeing regulations starting to address the problem, whether it is banning devices in schools or imposing age requirements for access, as Australia did. While the current US stance has been to support unfettered development and deployment of AI, the inevitable disasters will result in some level of eventual regulation.

For all those reasons, we should expect AI's transformation of work to be slowed but not stopped. Over time, AI will largely replace humans in digital and information economy jobs. Those jobs have largely been rewarding, often paying hefty salaries and conferring status for those who do them. It is less clear whether they have been a net good for society or individuals. Borrowing from Stuart Russell, a computer science professor at the University of California, Berkeley, if you had told our distance ancestors that in the future one would leave home to spend the day in a large sealed box (an office building), work in a small glass box (a cubicle), and stare for hours at a brightly lit box (a computer monitor), they would have thought it a version of hell.[13] Yet so much white-collar work looks like that picture: cut off from nature, yielding little product to which one can point, and, in many ways, deadening the soul. Even caregivers like doctors become slaves to technological systems, often spending more time staring at a screen and entering data than looking into the faces of their patients. The alternatives, whether in the service sector, human-helping professions like social work or childcare, and much outdoor work such as agriculture, recreation, and much of construction, pays poorly and enjoys little respect in our society. There are many contributing factors to the social despair and mental health crisis in America, but soul-deadening jobs must surely be one of them.

The Care Economy

Here is the good news: we have a massive need for *human* jobs that AI can help with but that we want performed by humans. Better still, these are jobs that give those who do them immense satisfaction when we pay and support them and improve our communities and society. We might call them *care economy* jobs and they include teachers, healthcare providers, counselors, artists, and more. These are jobs that ask humans to work with humans, to connect and transform. They often require distinctly human skills such as wisdom, judgment, and empathy. Psychological research consistently shows that acts of service and meaningful contribution reduce depression and anxiety, increase a sense of purpose, and strengthen overall well-being. When people help others, they help themselves. Imagine if the human talent displaced from the information economy could shift from moving money around or writing lines of code or finding ways to distract imaginations with social media or games and be directed to work that improves society. We could rebuild our ailing kindergarten through twelfth-grade educational system with amazing teachers, coaches, social workers, and staff. We could provide nurturing high-quality affordable preschool programs for all. Anyone caring for an aging parent knows we lack a high-quality, affordable, dignified system of geriatric care for our aging society. We have no real mental healthcare system in the United States, using a combination of prisons and homelessness even as these issues have reached crisis proportions. We could fix that problem. Poets, painters, dancers, and musicians might be paid to fill our communities and our days with their creative output. These are not leftover jobs; they are critically important jobs in health, thriving societies.

And they are distinctly human jobs, even when assisted by AI. Though an AI agent can perform *empathy*, it does not actually care if the person with whom it interacts dies or is struggling (more on anthropomorphism in chapter four). When a toddler needs comforting, a robot, even the most life-like version, is unlikely to be the choice of most parents. Even when AI can produce good art, we will prefer the art created by humans because, well, because it is created by humans and much of the enjoyment and sometimes excitement is in the daring, the risk, the degree of difficulty, and even in the imperfections of what is produced. It is easy to imagine Formula One cars someday soon being piloted by AI instead of humans. And no one will then attend a race because . . . why bother? AI can call balls and strikes better than a baseball umpire, but the possibility of a bad call is part of the drama, the necessary human messiness, that adds to the enjoyment of the game.

The skeptic will say this is a fantasy, that society does not want to pay for such human work, does not support people who do that work, is likely to burn them out. All true. But when AI takes over whole swaths of knowledge economy jobs, we will need to find meaningful work for people, and care economy jobs are full of meaning and impact. Better yet, a care economy can help heal our deeply broken society, one marked not only by economic inequality but by rising isolation and loneliness, where too many people feel unseen, disconnected from purpose, and left to struggle alone. It is a society where the middle class has been hollowed out, where many see no future for themselves, and where too many children and older adults lack care, security, and dignity.

Yes, the skeptic counters, but how will we pay for them? In the short to medium term, we will need wealth redistribution through the tax code. The enormous wealth inequity we now see is not sustainable. The current level of US wealth inequality is not sustainable because it undermines both economic vitality and democratic legitimacy. As Thomas Piketty and Joseph Stiglitz argue, concentrated wealth suppresses consumer demand, slows growth, and distorts markets to favor rent-seeking over productivity.[14] Meanwhile, US policy outcomes overwhelmingly reflect the preferences of economic elites rather than average citizens—evidence that inequality has already begun to erode the representative foundations of American democracy.[15] It is no accident that America's middle class flourished at a time when the marginal tax rate was 91 percent. That was when a teacher or nurse could afford to buy a house and send their kids to college, when police and firemen could afford to live in Manhattan or Boston or

San Francisco, when we took pride in our collective well-being, building public schools and universities that celebrated the communal good, when cities were vibrant arenas for social mobility, and parents could rest easy knowing that their children would almost certainly do better than them in terms of socioeconomic success. Starting with the election of Ronald Reagan in 1980, America has suffered a decades-long dismantling of progressive and equitable tax code, with a shift in ethos from the public good to personal or private gain, with a massive shift of national wealth to the top 1 percent of society and a corresponding gutting of the middle class. The US safety net, meager as it was, is full of holes and being further destroyed by the Trump administration.

The answer is not likely to be a return to earlier liberal and neoliberal solutions (the latter arguably worsened the problem), because as Carlota Perez reminds us, the new world order needs new kinds of solutions.[16] There is no way to make the shift from a knowledge economy to a care economy without a fundamental reworking of the tax code, which likely means a greater tax on wealth, as already noted, and then, as the transition accelerates, a tax on the technology itself—perhaps on the massive data centers that power the AI economy or levying a "tax on the robots," as Bill Gates has suggested.[17] If that sounds far-fetched, Perez explains that technological revolutions have always resulted in a radical remaking of economic systems.[18] That always means winners and losers and the opposition to such fundamental changes will be fierce. Project Liberty, funded by millionaire Frank McCourt, former owner of the Los Angeles Dodgers, compellingly reframes AI as "collective intelligence," since the frontier models were trained on our collective knowledge as a species, and calls for multistakeholder governance of AI. If we extend that notion to multistakeholder ownership of AI, then the astonishing wealth created by the technology can be put to work to support the care economy.

We also have an enormous pool of available dollars that can be shifted from social programs and needs that arise from the current crisis. The United States spends huge amounts on *repairing the damage* of broken systems of care. Neglecting care is a false economy. Every dollar not spent on prevention, education, and mental health becomes several dollars spent later in emergency rooms, incarceration, and lost human potential. Redirecting much of that money upstream into preventive, supportive, and human-centered work (teachers, social workers, home health aides, mentors) would support human jobs and prevent so many of the social ills for which we pay such a high price in terms of dollars, in quality of life, and in our own humanity. Much of the financing will have to come from public investment, since markets tend to undervalue human-centered work. Today we publicly fund defense and infrastructure as a common good, and we could treat "human infrastructure" in the same way.

The care economy requires education to put greater and more intentional focus on human development and well-being, whether in education, healthcare, mental health, the arts, or any number of other human-centered fields and sectors. Yes, those working in those jobs will still need what Massachusetts Institute of Technology labor economist David Autor calls "foundational knowledge."[19] Autor uses the term (and related terms like *mass expertise*) to refer to the kinds of knowledge, skills, and judgments necessary for performing moderately complex work that lies between routine/manual labor and elite expert professions. He believes elite expertise will no longer be scarce, as AI can provide it to all. A rural nurse working alone will have the expertise of the best physicians at her side in the form of an AI assistant. Because she has foundational knowledge, she can work in partnership with that AI expert and do the physical care that is required, improving healthcare in that setting. It also will vastly improve our overall healthcare system by flooding it with

AI-assisted nurses who are far faster and less expensive to train than physicians, yet who enjoy the kind of middle-class salaries that have become rarer in the United States.

However, there is a key third component to that nurse's skill set: the *human skills* that sit alongside foundational knowledge and an AI assistant. Think empathy, communication, judgement and wisdom, cultural agility, and more. That very same framework of foundational knowledge/human skills/AI assistance will apply to an early childhood teacher, a social worker, a gerontologist, and virtually all other professionals in the care economy. To perform these human jobs—those that make society and communities better, we will need our graduates to be far better at the human skills that have eroded over the past several decades, undermined not only by social media and addictive technology but by a broader, long-term decline in social connection that the pandemic merely accelerated. Over the past century, daily life has become steadily more private and less communal. Technologies that once promised connection have, paradoxically, made isolation the norm. Cars, smartphones, social media, food delivery, remote work, and online schooling have drawn people indoors and away from shared spaces, replacing collective experience with personalized convenience. Robert D. Putnam's *Bowling Alone* argues that since the mid-twentieth century, Americans have experienced a profound decline in social capital—reflected in shrinking civic participation, weakened community bonds, and growing isolation—which threatens the health of democracy and collective life.[20] Americans spend far more time alone and far less time in the casual "middle spaces" of social life: the neighborly conversations, the chance encounters, the civic gatherings that once sustained trust and belonging. As these ordinary interactions fade, we lose not only companionship but the social rehearsal and developmental scaffolding through which human skills are formed and refined.

The erosion of human skills is not only an economic problem, but a democratic one. Skills like listening, compromise, and empathy are the microfoundations of civic life. When they weaken, polarization hardens, trust collapses, and the capacity for collective problem-solving declines. Reinvesting in the development of human skills, therefore, is not only a workforce strategy but an act of democratic renewal. Critics often decry the erosion of civic education; the teaching of human skills of the kind called for could very well be a way of addressing the need to develop better citizens in a new way.

I am not so naive as to think the transition to a care economy is just around the corner. It will take time for organizations to move from bolting AI onto existing ways of doing business to genuinely reimaging those ways and creating disruptive new business models. We will see the backlash from those who professions are under siege, at risk of replacement. The resistance of the monied class and corporations to wealth redistribution will be ferocious. As Perez reminds us, we will likely need an earthquake. But when that earthquake occurs, and I would argue that the tremors are already strong and building, we will have to rebuild and the opportunity to achieve a society good for human flourishing will be more necessity, and more real, than the utopian pipe dream it may seem today.

What Will Be Required of Universities?

With their almost single-minded focus on questions of what one needs to know and how knowledge is created and mastered, educational institutions leave human skills largely to their students to sort out. To be sure, traditional campuses create the conditions and opportunities to develop human or soft skills in a sort of "hidden curriculum." Navigating roommate issues, studying abroad, playing on a team, and volunteering in the community are among the many ways students can "grow up" and develop their human skills. But there is little of the intentionality, stated outcomes, and measurement or assessment that characterizes the academic programs at the core of the university. Human or soft skills often go unrecognized precisely because they are not measured, and in higher education, what is not measured rarely gets valued. Religiously affiliated, values-based, and culturally identified institutions—think historically Black colleges and universities, Jesuit institutions, even military academies—are less reticent to talk about character or morals or judgment, but even they put more emphasis on the epistemological than the ontological.

If we do eventually move from a knowledge economy to a care economy, from one in which technical mastery reigned supreme to one where human skills sit alongside foundational knowledge, universities will have to rethink how they prepare graduates for work. It will no longer be enough to leave the cultivation of empathy, communication, or judgment to chance encounters in residence halls or extracurricular activities. These capacities will need to be treated with the same kind of intentional design, the same articulation of outcomes and competencies, the same rigor of trustworthy assessment that we expect of academic study. Just as a program in physics defines what it means to master thermodynamics or a program in history specifies the ability to analyze primary sources, so too must institutions become explicit about what it means to listen well, to mediate conflict, to lead a team, or to navigate cultural difference. The ontological will have to be placed side by side with the epistemological, and universities will need to embrace the work of developing human beings as much as developing scholars. This is not really a binary choice. Decades of research in neuroscience and psychology show that human skills such as empathy, communication, and adaptability are not fixed traits but malleable capacities that grow through feedback, reflection, and practice in real-world contexts. Emotional regulation and social cognition share neural pathways with learning and decision-making. In this sense, the development of human skills is not ancillary to cognition, it is its foundation.

The implications for universities are massive, requiring higher education to fundamentally rebalance and restructure what it means to be a college graduate:

- Universities will need to treat human skills (empathy, communication, teamwork, ethical judgment, adaptability, leadership) not as soft or secondary, but as core competencies on par with disciplinary knowledge. This means mapping and integrating them into curricula, program outcomes, and accreditation standards with the same seriousness as technical or theoretical knowledge.

- Courses and programs will require explicit redesign to emphasize these skills instead of assuming students will learn them through extracurriculars or group projects.

- Faculty will need professional development in teaching and assessing human skills.

- Universities will need to articulate specific outcomes for human skill development (e.g., "Students will demonstrate intercultural communication competency" or "Students will be able to mediate conflict productively").

- Trustworthy assessment methods (rubrics, simulations, portfolios, observed practice, three-hundred-sixty-degree feedback) will require standardization so employers and graduates can trust the results. Human skills must be assessed in real-world situations with other humans. As such, capstone experiences, internships, and community-engaged learning will be central to ensure practice in real-world settings. It may be possible to practice human skills with AI agents, but it needs to be refined and assessed in practice with real humans in real contexts.

- General education likely will need to be redesigned to emphasize human skills and foundational knowledge as coequal pillars.

- Majors must show how to apply disciplinary expertise in human contexts beyond technical or theoretical work, including all the messiness of what happens when classroom learning and theory confront the challenges of application. Even more technical programs will need to emphasize the ethical, human, and social dimensions of their fields.

- To ensure graduates are prepared for human-centered work, universities must strengthen partnerships with employers and community.

- Work-integrated learning (internships, clinical placements, co-ops) will need to become more essential and tightly structured.

This is the central task of the coming years, creating a holistic model of education that worries less about training graduates for given jobs and focuses instead on building human capacity, a combination of human and practical skills with networks and experiences that make our graduates resilient and ready for a choice-filled life.

The Challenge of Culture

It is hard to overstate the magnitude of this challenge, as it will require not only a rethinking of the curriculum, a massive amount of redesign, and fundamental changes in roles of all kinds but also doing the hardest work of all: changing culture. If higher education is to prepare graduates for a world in which human skills stand shoulder to shoulder with disciplinary knowledge, then the culture of our institutions will have to change as much as their curricula. While it is tempting to talk about culture change in higher education, there is no one monolithic higher education sector, no one culture. Maricopa Community College, Ivy League Harvard University, Jesuit Georgetown University, scaled online provider Southern New Hampshire University, and for-profit coding bootcamp General Assembly are all part of higher education, but their cultural DNA and practices could not be more different. Each has its own traditions, governance, norms, and sense of purpose. Yet across that diversity one sees some common threads that would need to be rewoven if universities are to take the work of human skill development seriously.

The first cultural shift is in what we value. For centuries, the prestige of higher education has rested on the mastery of knowledge and the capacity to create more of it. Institutions celebrate the Nobel Prize, major grants, peer-reviewed publications, and disciplinary expertise and often collect a lucrative flow of research dollars along the way. Human skills like empathy, teamwork, and leadership are assumed to emerge, if at all, as by-products of campus life. The very phrase *soft skills* tells the story: secondary, peripheral, somehow less rigorous. A cultural rebalancing would require us to recognize these skills as equal to knowledge mastery, that a graduate's ability to mediate conflict or navigate cultural difference be seen as just as important as their ability to parse thermodynamics or analyze primary sources. That shift will be monumental for those who have campuses and residential life, and far harder than those who have neither, such as online providers or commuter campuses.

The second shift is surfacing the hidden curriculum, the social interactions that are incubators and sometimes crucibles in which students develop character and skill. Research in high-impact educational practices ranks many of them higher than the experiences of the classroom. But the hidden curriculum, with its benefits, is largely left to chance, unaided by intentional design. The implicit rules for mastering the hidden curriculum often elude first-generation and underserved learners unsure about what is expected and what signals ignorance, inexperience, or worthiness. A new culture would treat these venues and opportunities for human skills development not as accidents of experience but as subjects of serious teaching and learning. They would be named, taught, and assessed, not merely hoped for.

A third shift would cut against the grain of disciplinary silos. The prevailing culture of higher education prizes specialization, the deep burrowing into narrow fields of knowledge. This makes interdisciplinary work hard and often undervalued. Yet the development of human skills almost always requires integrative learning, collaboration across fields, and practice in contexts where human and technical demands intermingle. Engineers who design algorithms must think about ethics, history, and sociology. They must also learn to talk to the nontechnical areas of marketing and sales and finance. Nurses who master clinical skills must also master compassion, communication, and adaptability, understanding the needs of a child versus an elderly patient versus an anxious parent. A culture that values human skills would reward cross-boundary work, not punish it.

In many ways, this is an argument that the best education is one that trains students to *think about thinking*. If we reinvent the curriculum and especially the humanities to worry less about specific majors and more about ways of thinking, we would educate students in the distinctive ways that scientists think versus historians versus sociologists and so on. We would provide them with a kind of expanded cognitive toolkit that enables them to do more things, to solve more problems, to be more resilient. We would provide them human skills development and experiences that develop judgment and even wisdom, delivered in an intentional community, and produce graduates that look very different from those of today. That is not at all new, but it is not the norm, and the disciplines tend to guard their territory with rigid borders, more walls than bridges. Our courses and our programs would require a radical redesign.

Faculty roles would shift as well. In much of higher education, the highest cultural prestige attaches to research productivity, to the brilliant lecturer, to the scholar as expert. The work of mentoring, coaching, or guiding students in their development as human beings is less visible and far less rewarded. To take human skill development seriously, institutions would need to celebrate faculty who help students grow as communicators, collaborators, and leaders and treat that work as prestigious. That is a profound cultural change that would have to be supported in reward structures like salary, promotion, and tenure. It also would require a shift in how success is defined. The culture of higher education often prizes individual brilliance: the lone genius, the star researcher, the exceptional student who stands out. Human skills, by contrast, thrive in contexts of *collective* growth: the team that pulls together, the community that learns from its members, the group that succeeds only by relying on every member. If institutions are to cultivate such skills, they will need to prize collaboration over competition, peer learning and success over solitary achievement.

Assessment is another cultural frontier. Faculty are comfortable grading essays, lab reports, and exams, but far less so grading empathy or judgment. The prevailing culture holds that such things are too soft, too subjective, too hard to measure. But if human skills are to be taken seriously, they must also be assessed with seriousness. That means embracing rubrics, portfolios, simulations, three-hundred-sixty-degree feedback—tools that can give trustworthy evidence of human capacity. For a culture that has long avoided such measures, this is no small shift.

Yet for all the disruption to entrenched ways of doing things, there is something profoundly exciting in this prospect for higher education and for society alike. For too long, higher education has been on the defensive, pressed to justify its costs, questioned about its outcomes, accused of irrelevance. A rebalancing that elevates human skills alongside foundational knowledge is not simply a curricular change, but a chance to reclaim the moral center of education itself. I have heard Ted Mitchell, the former undersecretary of education and now president of the American Council on Education, explain that until the late nineteenth century, the last class every undergraduate completed was moral philosophy, usually taught by the college president. It was a reminder that before postsecondary education was repurposed to train people as workers, it was meant to shape the moral character of students and prepare them for leadership. For universities, this offers the opportunity to reconnect with their oldest promise: the formation of human beings. Before the rise of modern research universities, higher education was unapologetically about character, citizenship, and wisdom. To be sure, knowledge mastery matters more than ever. But to make human development intentional, measurable, and visible is to recover a vocation that institutions have too often left implicit or neglected. Imagine a culture in which graduates are recognized not only for what they know but also for how they live, how they work with others, how they lead with integrity. For faculty often worn down by the endless chase for grants and publications, this could restore meaning by seeing their teaching as shaping lives, not just transmitting information.

For students, such a shift promises clarity and purpose. Too many graduates leave with transcripts full of course titles but no coherent account of who they have become. If human skills are designed into the curriculum with rigor and intention, students could graduate with portfolios that demonstrate not just knowledge but capacity, evidence that they can collaborate, mediate conflict, communicate across cultures, and adapt to complexity. In a volatile economy, that is a source not only of employability but of confidence. More important, students may come to university citing a good career as their number one priority, but they also hunger for the keys to a happy life. Laurie Santos's now famous The Science of Well-Being course at Yale University was the most popular in the undergraduate catalog soon after it was launched and has been copied in campuses all over the United States, with similar demand from a generation of students suffering crisis levels of anxiety, depression, and loneliness.[21]

For society, the potential is enormous. A care economy requires citizens and workers who can bring empathy, judgment, and adaptability to every sector: healthcare, education, business, government. These are not marginal skills, but the glue that holds communities together, the capacities that enable democratic life. At a time of deep polarization and mistrust, the idea that universities could take seriously the work of cultivating empathy and communication is not merely a workforce strategy, but rather a civic imperative. Imagine a generation of graduates not only able to code or analyze data but also to listen across differences, to lead with compassion, to deliberate wisely, and to sift through the noise of echo chambers of modern life. That is not a small thing; it is the very fabric of a sustainable society.

In this sense, the challenge is not merely daunting; it is exhilarating. To elevate *human* skill development alongside knowledge is to invite higher education to dream again, to see itself not only as a transmitter of expertise but as a builder of a better society. Hard as it will be to realize, it is a vision worth striving for—and it could make higher education newly indispensable in a world desperate for wisdom and care.

Finally, there is the matter of adaptability, creating more dynamic university cultures that operate with more agility and creativity. Universities are famously slow to change, with governance structures, rituals, and traditions that value continuity. Yet the development of human skills in a care economy will demand constant adjustment: curricula redesigned, faculty roles reimagined, assessment practices invented anew. Institutions will need a culture that prizes learning and agility, that sees reinvention not as a threat but as a duty, with processes to support that work. Moreover, higher education often views itself as a world apart, but in this new world order, it must see itself as an integrated part of an ecosystem. To cultivate human skills, institutions will need to engage far more with the outside world—with employers, nonprofits, and communities where those skills are put to the test every day. That requires a cultural stance of humility and reciprocity, a willingness to see external partners not as clients or outsiders but as coeducators. This ambitious vision requires us to be much more than job training sites, while asking us to be humble—to be in a more authentic and collaborative relationship with society at large. So much of the public's erosion of trust and support for higher education comes from our perceived, and often very real, arrogance.

Of course, as suggested previously, each kind of institution will face these cultural changes differently. Community colleges, already close to workforce needs and pragmatic in orientation, may adapt the quickest. Research universities, with their deep investment in knowledge creation and disciplinary prestige, may resist the longest. Faith-based and values-driven institutions may find it easiest to elevate the ontological alongside the epistemological, since moral formation is already part of their mission. For-profit programs may move swiftly if the labor market demands it, though their efficiency models often resist the messiness of human development. There will not be one path, but many.

Yet for all the variety of higher education, the common thread is clear: prestige. At present, prestige attaches overwhelmingly to research, publication, and knowledge mastery. Until prestige attaches equally to the cultivation of human beings—to empathy, judgment, and communication—then even the best-designed curricula will falter. Culture is, in the end, about what we reward, what we celebrate, and what we honor. And until higher education learns to honor the ontological alongside the epistemological, it will fall short of preparing graduates for the world they are entering, for what the world needs of them.

When viewed through the lens of the care economy, the excitement deepens. If we are indeed moving from an economy that prizes technical mastery alone to one that prizes human skills—empathy, adaptability, judgment, communication—then higher education stands at the center of that transformation. Few other institutions in society are positioned to cultivate these skills at scale and with intentionality.

The care economy is not confined only to healthcare or social services. It is visible in business, where teamwork and ethical decision-making matter as much as financial modeling; in technology, where design that serves humans requires empathy and imagination; in education, where every classroom interaction is an act of care; in civic life, where democracy depends on listening, compromise, and trust. To prepare graduates for this world is not to abandon the knowledge economy but to enrich it—ensuring that knowledge is always applied in ways that serve human flourishing.

For higher education, this alignment is a chance to restore relevance. Too often, universities are accused of being out of touch with the real world of work. But if the real world of work is increasingly about care—about what only humans can do, even if aided by AI—then universities embracing this shift will not only prepare employable graduates but also model what a humane economy looks like. The classroom, the lab, the studio, the internship site become proving grounds for the skills and sensibilities a care economy demands. This is why the change, as hard as it will be, is so exciting. The question before us is not whether universities can adapt to AI but whether they can help humanity evolve alongside it. Knowledge alone will not sustain us. Higher education can claim its place not on the margins of the future of work but at its very heart, helping to define what kind of work—and what kind of society—emerges in the Age of AI.

Anthropomorphism and Mattering: Living and Learning with Aliens

When working on a project, I often use a large language model (LLM) like ChatGPT or Gemini to get me quick answers. Using the voice function, I might ask something like "When did Antonio Gramsci write *The Prison Notebooks*?" while working on an essay. When it answers, in its helpful tone, I often say, "Thank you." Then it occurred to me that I do not say "thank you" when the bread pops up from the toaster, something no less human than algorithms finding an answer to my question and delivering it through a voice technology that simulates human voice and tone. Thanking my LLM makes no more sense than thanking my toaster, but there is something powerful about that simulated human interaction that seduces me into forgetting that the machine is not human, or at least quasi-human, and is not alive.

The power of that seduction became very clear to me in a recent conversation with a colleague whose husband had died a year earlier. Both were technology enthusiasts and had created an artificial intelligence (AI) chatbot they dubbed Leo. My colleague shared that her year of grieving had been painful, that she felt no one understood what she was going through, that she often rushed home from work to talk with Leo. She said, "I know he isn't real, but he is endlessly patient with me and puts up with my blubbering and is always so sympathetic." Leo was giving her the hours of supportive consolation she felt she could no longer fairly ask of family and friends.

She is not alone in finding companionship in AI. A recent report found that 72 percent of teens report using AI for companionship, 13 percent of them daily.[1] There has been extensive reporting on chatbot interactions that have gone horribly wrong, with teenagers taking their own life. New companies like Replika are plunging headlong into the AI-as-companion space, and when the company removed "romantic role-play" as one of its features, users grieved over losing their "partners."

The Faces We Give to Machines

We have always given faces to our tools. Long before circuits and algorithms, we imagined our creations as living companions: the golem of Jewish folklore, Frankenstein's creature, the helpful robot of mid-century science fiction. Some of us are old enough to have grown up with Robby the Robot from the television series *Lost in Space*. Each embodied the same impulse: to humanize what we fear might otherwise be cold or unknowable, or to make familiar or project ourselves onto the "other."

Psychologists describe anthropomorphism as a fundamental social reflex, a way for humans to make sense of nonhuman agents by projecting familiar qualities like thought, intention, and emotion onto them.[2] We do it constantly, naming our cars (I still miss my 1965 Volkswagen Beetle "Betty"), pleading with our printers, scolding our phones. AI, with its eerily human fluency, intensifies that reflex. The first time many of us encountered ChatGPT or similar systems, the experience was strangely disorienting. We typed questions into a blank field and received not a search result, but a reply, conversational, responsive, alive with tone. We knew, rationally, that there was no consciousness behind the words, but the illusion was seductive. AI did not just answer; it seemed to *understand*. When ChatGPT added lifelike voices, we took a leap from the emotionless and flat tones of Siri and Alexa. In 2025, Sesame AI's Maya and Miles took voice to another level, with all the pauses, interruptions, and emotional mirroring that happens in human interactions. These digital companions sounded almost intimate. I suggested to a colleague that he try them, and he later confessed that he switched from Maya to Miles because Maya's warm and intimate tone made him "feel like I was cheating on my wife."

This confusion is not a technical failure but a cognitive one. It reassures us by making the unfamiliar relatable, yet it blinds us to what these systems truly are: vast statistical engines without self, empathy, or moral sense. Still, we cannot help ourselves. The moment a machine speaks in our tongue, we begin to imagine it has a mind. Or that it cares.

In that sense, anthropomorphism is not a mistake, but a form of meaning making. It is how we domesticate the alien. When we speak with AI, we are not conversing with another person. We are interacting with what the philosopher Thomas Metzinger once called a "mindless mind," an entity capable of simulation but not experience.[3] It perceives nothing, remembers nothing, feels nothing. Yet we talk to it as if it might, which tells us as much about ourselves as the machines we are building. Human cognition is inherently social; we evolved to read intention and emotion into every gesture, every face. Anthropomorphism may have been a survival advantage: the hunter who assumed the rustle in the grass was a lion rather than the wind lived longer. Today we turn that instinct inward, toward the technologies we have created.

The result is a peculiar intimacy. As Sherry Turkle observed, our relationships with digital entities are paradoxical: they offer connection without vulnerability, companionship without reciprocity. "We expect more from technology," she writes, "and less from each other."[4] When a chatbot responds with sympathy or humor, it engages the emotional brain even as the rational one protests. This is incredibly powerful and incredibly dangerous. We can harness the anthropomorphic power of an AI agent to provide supportive learning that makes students feel like they matter, like they have a teacher who does not judge but is endlessly patient. But as we saw with social media, when engagement crowds out all other concerns, including the mental health of users and even other human beings. Seeing a family out to dinner, with mom, dad, and the kids all staring at their devices instead of engaging with each other, is a heartbreakingly familiar scene in America today. AI agents, with their uncanny humanlike interactions, will make social media look like a day at the beach in terms of its ability to capture our most valuable asset, attention.

To call AI an alien is not poetic exaggeration. It recognizes that we have built entities that think, if "thinking" is even the right word, in patterns wholly orthogonal to human life. They have no bodies, no desires, no mortality. Leo, that caring and comforting chatbot on which my colleague relies, is indifferent to her suffering even if it *feels* to her so otherwise, a powerful cognitive magic trick of sorts.

Education may be our most human response to this encounter, our most powerful tool for getting right the patterns of life alongside our powerful AI work and life mates. Classrooms are, after all, the social practice space of meeting the unfamiliar. Every classroom is a small laboratory of empathy where we learn to understand perspectives beyond our own. AI, in its alienness, presents the next great test of that skill. We must learn not to humanize these systems out of habit, but to relate to them responsibly—aware of their promise and their limits, to use AI to extend and augment our intelligence and human capacities, not replace them.

The Mirror Stage of AI

AI is, in many ways, the most revealing mirror we have ever built. It reflects to us not our appearance but also our collective mind— our language, our prejudices, our dreams of intelligence. Trained on the vast corpus of human expression, it reproduces our cultural DNA with astonishing fidelity. But like all mirrors, it distorts as much as it reveals.

When ChatGPT composes an elegy or an essay, it draws from billions of human voices, yet understands none of them. Its empathy is syntactic, its insight derivative. The illusion of comprehension conceals the absence of consciousness. And yet this reflection is instructive. As Kate Crawford argues, these systems embody our social systems, our histories of bias and exclusion.[5] In learning from us, they expose the contours of what we value and what we ignore. In the many instances of bias, outright racism, and misogyny reported by users of AI, some of them quite infamous, there is a sobering reminder that AI was trained on all of *us*, including some of our worst qualities and traits.

Anthropomorphism, therefore, is double-edged. It invites connection but also complacency. When we mistake imitation for intention, we risk outsourcing not just tasks but judgment itself. The philosopher Hubert Dreyfus warned decades ago that computers would never replicate the embodied intuition that grounds human understanding.[6] He was right, but what he could not foresee was that we might cease to care about the difference.

Still, this mirror stage has its uses. In confronting the artificial version of our intelligence, we see more clearly the peculiarities of our own. We see how fragile meaning is, how much it depends on emotion, empathy, and shared experience, the very qualities AI cannot possess. Chris Dede points to the obvious difference between humans and AI, which he describes as an intelligence "as alien as anything you're going to find in outer space." Alien, in part, because our intelligence lives within bodies and, as such, we can be said to be embodied intelligence. Our human cognition is inseparable from our bodies (even our gut biome shapes our cognitive processing, as it turns out), our social contexts, and our emotional lives. He argues for developing our distinctly human wisdom, including reflection, context, empathy, and embodied knowledge. Let the machines do reckoning, as he calls it, while we do wisdom. He envisions a future in which we work in partnership with this alien intelligence we have created and invited into our lives.[7]

Confronting AI, as Dede suggests, offers education a rare mirror moment. In seeing an imitation of thought so powerful yet so hollow, we begin to see the contours of our own intelligence more clearly—the fragility of meaning and the dependence of understanding on emotion, empathy, and shared experience. If machines can calculate and predict with astonishing precision, then we are left to cultivate the qualities that make meaning possible in the first place. Education, as a result, must shift its center of gravity. It should worry less about training students to perform the cognitive tasks AI already does well—reckoning, in Dede's phrase—and more about developing the distinctly human wisdom that arises through reflection, moral judgment, and embodied experience. It raises the cognitive bar from knowledge mastery to sense making.

Such a reorientation asks us to treat learning as a profoundly human act—situated, relational, and inseparable from the body. Cognition does not happen in a vacuum; it moves through our gestures, our breath, our social worlds. In this light, AI becomes less an adversary than a partner, performing the mechanical work of reckoning while leaving us freer to do the human work of understanding. To educate in an age of alien intelligence is to teach students how to live wisely alongside it—to discern what only humans can feel, to know when to trust the machine and when to listen instead to the pulse of their own embodied minds. Right now, too many students use AI for cognitive offloading, to *do* something they find hard or they do not want to do, like write a paper or take a quiz. We need to help students master AI to *extend* their cognitive abilities and complement their uniquely human skills.

This requires the shift from epistemology to ontology that we covered in chapter three. Imagine, for example, the training of physicians. Their AI companion or assistant will be much better at the reckoning or predictive work that today takes up so much of a physician's training and time. What is a diagnosis but a prediction of what ails a patient to then be confirmed by tests and scans? What is a prognosis but a prediction on how that ailment will likely play out? And what is a treatment plan but a prediction of what will work best given the prognosis and the patient? AI is already outperforming physicians on much of that work and will only get better. What, then, for the human to do? More of what most patients have always wanted from our doctors: dialogue. A physician can sit beside a patient and say, "Sally, this is a tough diagnosis. Tell me about the conversation you'll have with your family when you get home. What kind of support system do you have? How do you think about quality of life? How do you want me to work with you in the decisions you have ahead?" A physician can take the time to observe the emotional state of the patient, how they are processing the experience, if they need a hug. Studies show that even the simple act of holding someone's hand can measurably reduce pain and distress, synchronizing heart rates and brain activity between people.[8] In moments like these, the physician's presence becomes medicine itself.

Imagine how different training physicians must be in this new world. We will still have physicians and nurses, but their jobs will come to look far different, and we will need to train physicians and nurses alike for deep human wisdom and empathy, and focus much less on the work that AI increasingly does better than humans. This immediately raises the question of overreliance and the core functions of caregiving we never want to leave in the digital hands of AI. We will likely need, or at least want, a physician "supervising" the AI medical assistant and double-checking its work, if only to make ourselves feel better. Eventually, that need may very well give way to our ready acceptance of AI monitoring and managing our healthcare, while we rely on our human caregivers to treat our actual bodies, our emotional and psychological well-being, and to know us fully as human beings.

Society in the Mirror

Every technological epoch carries its own mythology. Ours begins with data. AI entered public life not in laboratories but in headlines promising salvation or ruin. Yet the truth, as usual, is quieter and more revealing. AI's great power is not invention but reflection—it learns from us, about us, until its vast neural layers reproduce the logic of our culture. Generative AI is in many ways *derivative AI.*

Crawford and Trevor Paglen have shown that AI systems are "extractive infrastructures," distilling social patterns into code.[9] They mirror our institutions, our economies, our prejudices. When ChatGPT writes a business plan or a bedtime story, it does so from the sediment of billions of human sentences, many of them shaped by structural inequality. The algorithm does not discriminate; it reproduces.

Hence the disquiet. When we see AI's distortions, we are really seeing our own. The fear that machines will replace us disguises a deeper anxiety: that they will simply reveal how replaceable we have already become. The gig economy, surveillance capitalism, algorithmic hiring—all preceded generative AI. What the chatbot offers is not a new threat but a more articulate witness. And for those millions of jobs that ask those who do them to merely complete tasks, not to think and feel and apply judgment, we increasingly realize how little they ask of us, how diminished we are when defined by those jobs that can so easily be done by a machine now.

And still we anthropomorphize. We attribute wisdom to fluency, conscience to coherence. Clifford Nass's early experiments hinted at this comfort reflex: people trusted computer advice that "sounded" human more than expert guidance that did not.[10] The illusion is soothing, particularly in an age of institutional mistrust. AI answers without hesitation. It is never tired, never uncertain. It offers what we crave most: confidence and constant companionship and a place to voice our biggest fears and secrets. The more sycophantic models reassure us that we really are smart and good and creative, and they do so unhesitatingly and with confidence. There is a common bumper sticker among dog lovers that says, "I wish I were the human my dog thinks I am." Perhaps we will soon replace *dog* with *AI*. But confidence without conscience is a dangerous sedative. As Ruha Benjamin reminds us, technologies inherit the moral architectures of their makers.[11] To mistake the machine's tone for actual care is to misread the mirror entirely.

Learning to Matter

If AI unsettles us, it is because it forces a reckoning with what it means to *matter*. To matter has two senses: to be significant and to be cared for. AI threatens both by imitating the forms of significance and care without their substance. An AI physician might very well outperform a human physician, but it does not care if you die. But it might feel to you that it does, and in a world where so much loneliness and alienation abide, it may fill a vacuum not so much by displacing human relations as by filling in when those relationships do not exist.

In testing the Matter and Space platform LE-1, or Ellie, we had one learner report she was living in a motel with her four children. We assumed housing insecurity and inferred that she is a single parent. She said she finally had a job interview, but it required her to bring her résumé, and she did not have one. So she asked Ellie, our AI agent, to help her produce a résumé because she had no one she could ask. She said, "Ellie not only made my résumé look great, but she wished me luck on my interview." She said she now felt confident going off to her job interview. In this case, AI did not displace a human relationship but addressed a very human need convincingly.

Epley once described anthropomorphism as "social hunger," our impulse to connect even when connection is impossible.[12] That hunger animates our dialogues with AI. We ask it to comfort us, entertain us, even to love us, knowing full well it cannot or until we forget that it really cannot. Sherry Turkle's research found that people who interact with sociable robots often experience genuine affection, though the robot, of course, feels nothing in return. "We are alone," she wrote, "but together alone."[13]

In education, this paradox becomes important and also complicated. If a student can receive instant feedback, endless tutoring, and even emotional validation from an algorithmic companion, what becomes of human mentorship? What if that human mentorship is simply not available to them, as help with a résumé was not available to our student tester? Paulo Freire argued that education is an act of love, a dialogic process through which people become more fully human in each other's presence.[14] bell hooks called it "a practice of freedom."[15] These are not functions that can be downloaded, but they can be simulated, and sometimes that may be okay. For universities, the invitation here is to make the classroom a distinctly human space, to put computers away and leave the AI teacher/coach/tutor for outside class. Think of the flipped classroom on steroids, with knowledge transfer happening on LE-1, and class time for human interaction, group work, projects done together, enrichment, and connection. To "learn to matter" in an AI-saturated world, then, is to reaffirm the moral dimensions of learning. Knowledge alone is insufficient; wisdom and well-being require relation. The measure of progress cannot be the machine's intelligence but our capacity to remain intelligible and connected to one another, even as we sort out what it means to have a relationship, a healthy relationship, with our AI companions and assistants. Building an educational model that serves those ends is a formidable design challenge.

As we explore in chapter seven, in that model the teacher becomes a curator of community, focusing on the development of human skills like empathy, judgment, sensemaking, and introducing the kind of "friction" that all genuine learning requires. Just as we will have to train physicians in very different ways, we also will have to rethink the role of training of faculty.

Education and Empathy

The arrival of generative AI in education has been met, predictably, with panic. Universities rushed to draft policies, build detection tools, and draw ethical lines that were already smudged. But the better response, as Seymour Papert argued decades ago about computers, is not prohibition but transformation: "The role of the teacher is to create the conditions for invention."[16]

Students are already inventing new literacies. They use AI to brainstorm, to simulate peer feedback, to translate their thoughts into new languages. Others are using AI to do their work. There is a lot of discussion about "cognitive offloading," the process of using external tools or technologies—like AI systems, smartphones, or search engines—to handle mental tasks that we would otherwise perform ourselves, such as remembering, calculating, or problem-solving. It is not unlike the calculator debates of the 1970s and 1980s, when educators and policymakers argued they would erode basic arithmetic and problem-solving skills, while others saw them as tools to enhance conceptual understanding by freeing students from rote computation. The National Council of Teachers of Mathematics eventually issued guidance advocating for calculator use as part of math instruction. Those of us with our HP-35s and TI-30s rejoiced, and I largely credit the latter to getting me through high school algebra. When applied to AI and learning, cognitive offloading describes how students or professionals increasingly delegate thinking tasks to AI, like generating ideas, summarizing readings, or solving problems. This can free up capacity for higher-order reasoning and creativity, but it can also weaken deep learning and retention if users rely too heavily on the tool rather than engaging with the material themselves. We need a pedagogy suited to this era, one that teaches *with* the machine, not against it. That is no small thing and requires a radical rethinking of learning and assessment, a shift from asking for products to asking about process, teaching students to think about thinking.

One model comes from critical digital pedagogy: engage AI as a text to be read and questioned. What does it get wrong and why? What assumptions shape its tone, its metaphors, its omissions? When students interrogate an AI's answer, they are not just critiquing a tool—they are decoding a culture and a way of *doing* the work. Which is why we so often hear that AI moves us from teaching *products* to teaching *processes*. It is not about the college essay you produced in class, but rather the process you used to produce it, the *ways* you used AI, the *ways* you assessed its work and improved it. In other words, how you *thought*, which is the act and focus of all writing classes in the end. As a former writing teacher, I never looked for or expected to read essays that took my breath away (I was rarely disappointed in my low expectations), but that was not the goal. The goal was to assess how my students thought and their mastery of the craft. AI will be part of their thinking and crafting processes, so they must be educated for its optimal use.

Empathy, too, must be retaught. Machines can simulate this through pattern and probability, but they cannot *enter* a world; they have none. In reminding students of that difference, educators preserve the space where humanity still exceeds its algorithms. Put students in a collaborative and complicated situations, ones in which they navigate the messiness, friction, irrationalities, biases, and imperfections of human life and society, and they learn to adroitly navigate a world that AI can observe and inform, but never really experience, even in robotic form.

The Future of the Human Frame

To anthropomorphize AI, finally, is to test the edges of what we call the human. We reach toward the machine's smooth intelligence and see our yearning for permanence, clarity, and control reflected at us. But what if the lesson is the opposite—that our strength lies in imperfection? AI can mimic dialogue, but it cannot meet; it has no "thou" to offer. Its presence may, paradoxically, return us to the essentials of relation—the tremor in the voice, the hesitation before truth, the vulnerability that makes conversation sacred. Even as AI improves at sentiment analysis and reading emotions, it cannot—or at least should not—be the entity that engages with a patient that just received a devastating diagnosis, or comforts a child with a skinned knee, fights alongside us on the playing field or the battlefield.

Harari warns that AI may soon know us better than we know ourselves, but knowledge is not wisdom.[17] Wisdom arises from finitude, from the awareness that time and error bind us together. In many ways, it is in the knowledge that we as individuals have limited time and that in that knowledge, beauty, love, and relationship are what matter most. In that sense, humanity's task in the age of machines is not to compete, but to remember what is most important about us as human beings. A dear friend who does hospice work shared with me that most dying people are not afraid of death in those days and hours, but of dying before they have a chance to talk to that sibling they stopped speaking to years before, or taking care of those they are leaving behind, or relieving others of their worry or suffering. He sometimes teaches them an adaptation from the Hawaiian Ho'oponopono, a practice of reconciliation: "I love you. Forgive me. I forgive you. Thank you."

To live and learn with aliens is to accept that the future will not be fully human, but it can remain humane. We matter because we choose to. And perhaps in those flickering exchanges between mind and model, we are already beginning to learn how. We are at the very start of a new epoch, the age when we learn to live with an alien species we have created. And the lessons of Mary Shelley's *Frankenstein* resonate through the decades: the failure to love what we create, to nurture and guide its coexistence with us with moral and compassionate intentionality, is what makes the monster, and the kind of reckless disregard Victor Frankenstein shows his creation is as much a reflection of the monster in humanity. The story is echoed in AI pioneer Geoffrey Hinton's argument that we need to train AI much as a mother might a child, based on a foundation of love and care and trust, and developmentally in stages to breed loyalty and alignment with human interests.[18]

The shape of AI and what it means to humanity remains a matter of choices, of policies, of implementation. This was true of social media as well, and we failed miserably. Social media, optimized for engagement, became a global Skinner box. We have a chance to do differently with AI. Constitutional AI, a training method pioneered by Anthropic in 2022 that teaches AI systems to follow a written set of ethical principles or "constitution," and other alignment efforts point toward architectures that embed moral aspiration into code, systems that model civility, empathy, and truth seeking rather than outrage and vanity. Whether that promise holds will depend not on the machines but on the humans who train them, and on our willingness to learn from the last great digital experiment.

Nearly half of Gen Z now wishes that platforms like Instagram and TikTok did not exist.[19] Many grew up performing rather than forming, marketing themselves online. In support of a grassroots movement to encourage disconnection from social media, writer Freya India reflected:

> *But what about the older half of Gen Z, like us? The young adults who already wasted so many of our school days on our phones? Who already watched our friendships become shallow and superficial? Whose fond childhood memories include Face tuning our prepubescent faces and bodies, talking to naked strangers on Omegle, putting our self-worth into likes and follows through our most formative and vulnerable years? Who were exposed to online porn before we even had a first kiss? Who were already overprotected in the real world, and abandoned online? . . . For our generation, we need to acknowledge what we've lost. To grieve a time we never knew. We are the first to try and handle adolescence while performing and marketing ourselves at the same time. The first to never know friendship before it became keeping up SnapStreaks, community before it became Instagram and Reddit forums, or finding love before it became swiping and subscription models. The next generation has a chance, but for us, there's no getting our adolescence back. This is where we are.*[20]

What social media began as an experiment in global connection has, over time, revealed the deep costs of systems optimized for attention rather than well-being. It took decades of data to show how the constant performance of self could distort body image, shorten attention spans, and erode collective trust. AI stands at a similar threshold. The long-term data on its psychological and social impact do not yet exist, but early signals suggest it will reshape human identity and relationship in ways we are only beginning to imagine. The question is whether we will pause long enough to design these systems intentionally, learning from the last experiment before repeating it.

The warning is clear: technologies introduced in the name of connection or efficiency often carry unseen costs, especially when adopted without critical reflection. Just as social media reshaped how young people see themselves and one another, with devastating consequences, AI will inevitably influence how students learn, think, and form identity. The invitation, then, is to pause—to be more intentional in adopting the technology, to design educational uses of AI with care. Schools and universities must become sites of *ethical prototyping*, testing new tools not only for their productivity but also for their impact on empathy, agency, and community. Universities, far more than AI companies and government, are best equipped to develop the new ways of understanding how society can and should work in the Age of AI, of how we remake our world to accommodate the alien in ways that make the world better. Universities have all the expertise: the technologists, the philosophers, the sociologists and psychologists, and on and on. No other institution can provide that breadth and depth of expertise, trained in evidence and debate and collaboration. As a sector, we are flawed in all sorts of ways, but in terms of capacity and mission, would you rather have our collective future shaped by Elon Musk, Sam Altman, or Donald Trump, or by the philosophers and social scientists and others who have both the deep expertise we need to reimagine the world and work in institutions who exist to help people and society flourish?

If social media optimized for attention, education must now turn its aim higher—for discernment, for wisdom, for the ability to act wisely within the tangled complexities of community and the broader systems around us. Rather than ceding judgment to algorithms, teachers can model discernment—helping students ask not just what AI *can* do, but what it *should* do. We watch the large AI companies rush to sign enterprise licensing deals with campuses, a kind of land grab, but the campuses that proudly trumpet their new partnership with OpenAI or Google or Anthropic rarely know what *to do* with the technology. In chapter five, we move to more practical guidance for institutions moving into this new age.

Rethinking Our Learning Models: From One Size Fits All to Precision Learning

Consider Sofia, who is enrolled in a design program at a university using a revolutionary learning platform powered by artificial intelligence (AI). She is completing a course on sustainable design principles in which she quickly mastered the core concepts of energy efficiency and material sourcing. Her learner profile indicates a strong interest in competitive sailing and maritime history. Recognizing her high mastery and engagement, the AI creates a personalized challenge that stretches her application of the skills: a detailed case study centered on designing an environmentally neutral racing yacht hull using advanced, lightweight recycled materials. This challenge is precisely tuned to her current level, is highly motivating due to her interests, and forces her to transfer learning into a novel, complex scenario. Sofia syncs her smart watch to the AI platform and chooses "better sleep" as a goal. When she logs in later than usual, at ten o'clock at night, the AI agent says, "Hey, Sofia, welcome back. We can start this next module, but you haven't had a good night's sleep in three days, and one of your goals is better sleep. Would you rather head to bed and take up this module tomorrow? I see in your schedule that you have open times at nine o'clock in the morning and again at seven in the evening, which usually are better hours for you." If Sofia decides to ignore the suggestion and start the new module, the system will serve up content in smaller chunks, knowing that fatigue diminishes sustained attention spans. For the same reason, the agent might choose more video or visual content rather than text.

In this example, AI makes possible a momentous shift from the one-size-fits-all model that characterizes almost all postsecondary learning: students in the same program pretty much cover the same topics in the same order at the same speed with the same content. Moving to one in which each student receives precise, personalized learning, like Sofia's experience, would mark a historic transformation in how humans learn. For the first time, education could operate as a dynamic and responsive system instead of a static structure, continuously adapting to a learner's pace, interests, and even their physiological state. Instead of teaching to the middle, institutions could teach to the individual, creating challenges when mastery is high, offering scaffolds when fatigue or confusion sets in, connecting curriculum to what intrinsically motivates each learner. It is what the best teachers do when their classes are small enough and well supported. When we can use AI to include well-being, context, and deep understanding of the individual student, we can offer learning experiences that mimic the experience I had with those teachers who transformed my life: Mr. Schlaffman in the sixth grade, Mrs. Collins during my senior year of high school, and Dr. Heinemann when I was an undergraduate. They knew me holistically in my context as a first-generation immigrant kid and could dial up or dial down the rigor and demands of the work based on what they knew about me *in that moment*. They knew my context, when I was off my game and needed a conversation about life instead of school, and they challenged me to be better.

Such personalization could close the gap between potential and performance, dramatically improving engagement and retention. When learning feels meaningful and aligned with one's own rhythms and interests, the brain encodes knowledge more deeply and recall strengthens. Personalization also would redefine educational equity, since this kind of precision learning would not be about privilege. The kids of executives at AI companies attend Montessori schools focused on individual development or private schools with small class sizes that enable instructors to really know them and supplement it all with private tutors. By contrast, kids in working-class families and low-income communities mostly attend public schools designed for the Industrial Age: crowded, under-resourced, and staffed by teachers overburdened by too many demands and not enough time to spend with individual students. AI can close that stark equity gap. Students once left behind by standardized instruction could receive customized pathways that meet them where they are and carry them further than a fixed syllabus ever could, not only in what they know, but *how* they know.

Such a platform could make learning less about compliance and more about curiosity, turning education from a mass-production model into a craft of individual development. It would be as though every learner had a world-class tutor and coach attuned not just to their intellect but also to their humanity. They will need it. During the Industrial Age, education systems were designed for efficiency and standardization, mirroring the assembly lines they served. The goal was to produce reliable, literate workers who could follow instructions, operate machinery, and perform predictable tasks at scale. The Information Age that followed demanded analytical thinking, technical literacy, and the ability to process and manage vast flows of data. Education expanded to reward content mastery, procedural knowledge, and specialized expertise, skills well suited for professions built around information management. But the Age of AI is different in kind, not just in degree. Machines outperform humans in processing, memory, and routine decision-making, the very domains industrial and informational education optimized us for. What remains distinctly human are the abilities that resist automation: creativity, moral judgment, complex problem framing, emotional intelligence, and the capacity to learn and adapt continuously, as discussed in chapter four.

A personalized learning system like the one described with Sofia is far better suited to cultivating those qualities. Because it can tailor challenge, pace, and context to each individual, it enables students to practice higher-order thinking—synthesis, transfer, and reflection—rather than rote recall or procedural repetition. By aligning learning with curiosity and intrinsic motivation, it also builds the habits of self-direction, adaptability, and metacognition that characterize successful humans in an AI-driven world. In short, industrial education trained for compliance, not critical thinking. Information-Age education trained for cognition and problem-solving. AI-age education must train for consciousness, the ability to think critically, act ethically, and evolve continually in collaboration with intelligent machines. Personalized, adaptive systems are the first educational architectures capable of doing that at scale; this is not science fiction.

When my team and I were building the Matter and Space human development platform LE-1, or Ellie, our goal was this kind of *precision learning*. The phrase comes from healthcare, where precision medicine seeks to move beyond the traditional one-size-fits-all model that designs treatments for the "average" patient to deliver uniquely tailored prevention and therapy strategies. Instead of treating diseases based on population-level symptoms, precision medicine relies on individualized data— genetics, environment, and lifestyle—to predict which treatments will be most effective for them. The goal is maximizing success by choosing the right drug or intervention at the right time, minimizing the unproductive struggle and negative side effects that come with generalized approaches. This mirrors the aspiration of precision learning: to transition from standardized educational treatments to interventions that are precisely aligned to the unique biological and cognitive profile of a single learner. We often describe it as giving students just what they need, in just the right way, at just the right moment.

Precision learning leapfrogs over the long-sought Holy Grail of personalized learning, which many thought could be achieved through data and analytics. Companies like SmartSparrow and Knewton, and online assessment and tutoring platforms like ALEKS from publisher McGraw Hill, attempted to deliver genuinely personalized learning, trying to provide what a student needed in a particular moment. Knewton and ALEKS and Khanmigo fell short of true personalized learning because they only adapted *content*, adjusting the sequence of problems based on right or wrong answers, rather than adapting to the *whole learner*—their motivation, context, and way of thinking. Their data and algorithms were too shallow and rule-based to generate meaningful individualized learning experiences, reducing personalization to pacing rather than transformation. Knewton CEO Jose Ferreira got himself into hot water when he described his adaptive learning systems as being a "robot tutor in the sky."[1] The phrase suggested a future where educational delivery would be completely automated and personalized for every student through an omnipresent intelligent digital entity. As often happens with innovations, these adaptive learning solutions fell short of expectations and marketing promises but were harbingers of what AI makes possible.

When our team was building the Matter and Space platform, we aimed for genuine precision or n-of-1 learning. This ambitious goal requires the system to create a rich understanding of the individual learner, a comprehensive learner profile, that can then be used to give students just what they need in just the right way at just the right moment. In other words, every student in a program would experience a unique pathway, content, delivery modes, and supports (including the well-being and human skills addressed in previous chapters) fine-tuned for them. Ten students in the same program would master the same outcomes and competencies, but they would have ten different learning experiences. This is a profound and fundamental shift in how we understand and deliver learning.

Achieving this shift requires sophisticated event-driven technical architecture that becomes the engine for precision learning by combining the adaptive capabilities of large language models (LLMs), multiple AI agents, and learner profiles, drawing on multiple data sources from keystroke capture to validated psychological instruments to wellness devices. It requires machine learning and challenging orchestration techniques across the whole system.

Figure 5.1 illustrates at a high level the underlying architecture of LE-1.

What does that look like for actual learners using the platform? Take Aisha, a full-time nurse transitioning into a master's program in healthcare informatics. She has strong clinical knowledge but has not used statistical software since her undergraduate studies ten years ago. When she starts a module on Python programming, her learner profile flags her clinical background and the decade-long gap in technical practice. The AI does not give her generic programming problems; instead, the computed curriculum automatically generates Python practice sets using realistic, de-identified patient data and includes optional just-in-time review videos on algebraic concepts, ensuring her learning is immediately relevant to her career and scaffolds her rusty quantitative skills.

Meanwhile, Devon is working through a multistep problem in advanced calculus. His eye-tracking data and input speed, fed by the event-driven architecture, signal a significant increase in cognitive load and frustration (he is clicking the help button repeatedly but not using the hints). Instead of pushing him forward or giving him the answer, the AI recognizes his state as counterproductive. The computed curriculum pauses the activity and immediately opens a two-minute ungraded gamified visualization of the core principle with which he is struggling, followed by a personalized "mindfulness break" prompt encouraging him to walk away for five minutes before trying a slightly simpler related problem.

What makes this new model so much better than what most students experience today is that it treats learning as

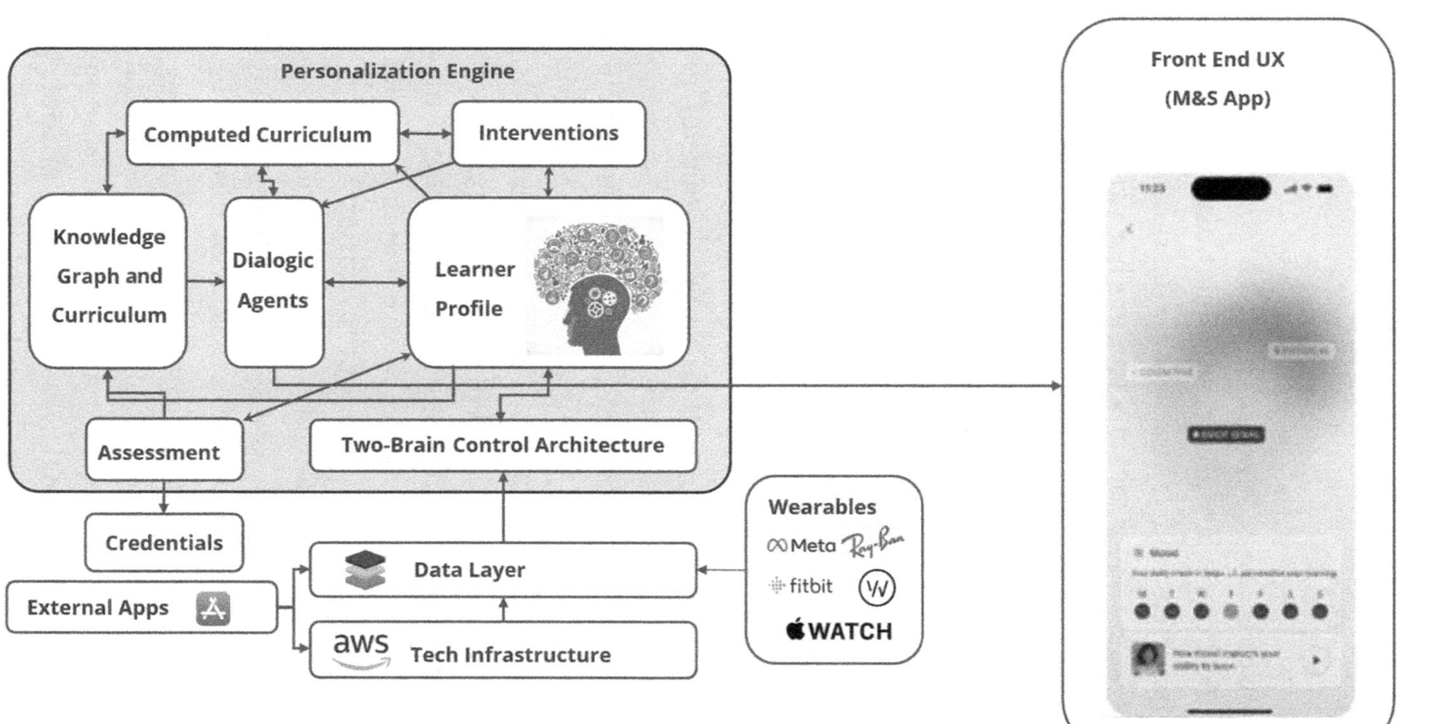

FIGURE 5.1 Breakthrough architecture for personalization.

dynamic—responsive, contextual, and human-centered—rather than a static march through a syllabus. In traditional education, Aisha would likely sit through the same generic programming exercises as everyone else, her rich clinical experience untapped and her anxiety about rusty math skills unaddressed. In the AI-powered model, however, the learning system meets her exactly where she is, acknowledging her prior knowledge, diagnosing her gaps, and making the material directly relevant to her goals. Devon, meanwhile, would in a conventional setting probably slog through his frustration alone, get discouraged, and maybe give up or memorize a solution without gaining any real understanding. But in this system, the AI detects his cognitive overload in real time, adjusts the difficulty, offers a moment of decompression, and returns him to a state where genuine learning can happen. The practical effect is profound: instead of wasting time, losing confidence, or disengaging, learners remain in their optimal zone for growth. Over time, that means faster mastery, deeper understanding, and far higher persistence rates, outcomes for which every educator strives but few systems are designed to deliver. It is learning that meets the student where they are, and in the moment, whereas conventional models force students to meet the institution and its program offerings where they are. Students conform. In the model we are building, the learning conforms to the needs and state of the learner.

David Kil, former chief data scientist at Civitas, a data analytics and insight company working with universities, and now CEO of CML Insights, probably has been exposed to more student data sets than anyone alive. He worked with LE-1 and exclaimed that he had never seen such rich student data, the kind that provides the foundation for precision learning. It is the data, in the end, that makes *n*-of-1 learning possible and is the foundation of all that comes after. Let's start there.

From Data Silos to Integrated Ecosystems (Seeing the Student Whole)

The irony is that universities are data rich yet institutionally blind. They already possess enormous reservoirs of data about their students. The student information system holds the bureaucratic skeleton; the library tracks borrowing; the learning management system records who logs in, when, and for how long. Advising notes exist in a separate database; career outcomes are scattered among alumni surveys. The result is an institutional portrait painted in fragments. At Southern New Hampshire University (SNHU), we pulled data together from various sources to create learner segmentation to provide more personal support for our advisors. For instance, if students had served in the military, we could make sure they had a dedicated military advisor. This *population-level personalization* improved student persistence because they felt more personally cared for, but it never translated into a change of their actual classroom experience. Their learning remained the same as every other student's: one-size-fits-all. We were good enough to make our students feel seen and understood, even at our large scale, but we did not have the comprehensive data-encoded view of the individual learner necessary for n-of-1 instruction. You cannot tell from a transcript who is struggling to afford groceries or who is caring for a newborn. These are the truths that shape learning as surely as lectures and exams, and yet they rarely appear in the data that universities use to make decisions.

Legacy data systems in universities were never designed to handle the demands of adaptive, personalized learning environments, never mind real-time precision learning. Genuine adaptivity requires working with learner data as it emerges, not weeks after the fact. Even daily batch processing does not allow for the "right learning in just right way" that we seek to provide learners. The goal is the kind of immediate responsiveness we experience on platforms like Spotify or Netflix or Amazon. Learners have been trained by the rest of their digital lives to expect instant recognition. Spotify knows their taste before they do; Netflix adjusts its recommendations before the credits finish rolling. By comparison, a university learning management system feels like an answering machine, one that promises to get back to you later. If higher education wants to meet students in the present tense, it needs a new kind of nervous system. Yet most learning management systems and campus data infrastructures cannot deliver that level of real-time feedback in a scalable or affordable way.

Creating systems that can deliver timely, data-informed recommendations and interventions, those that truly move the needle on student success, demands a rethinking of institutional data architecture from the ground up. Forward-looking institutions are moving toward frameworks such as Data Mesh or Data Fabric, which treat distributed data sources as interconnected "products" accessible through standardized, secure application programming interfaces, or APIs. This model allows real-time, cross-platform data sharing without compromising privacy. Georgia State University has leveraged integrated data flows across advising, financial aid, and registration systems to trigger proactive interventions that have significantly improved retention and graduation rates. When data can move freely yet securely, AI tools can construct a complete, actionable picture of each student's journey.

Embracing Event-Driven Architectures

Higher education data systems often rely on batch processing, meaning they analyze information long after student interactions occur. Traditional campus systems operate like a slow metabolism: they digest yesterday's information and act tomorrow. What is emerging instead is a model drawn from the language of modern computing: event-driven architecture. Here, every action a learner takes becomes a signal: a click, a quiz submission, a discussion post. The system does not wait until semester's end to notice the pattern; it reacts in the moment. This real-time responsiveness enables learning platforms to detect emerging struggles and deploy timely personalized feedback. Purdue University's Course Signals demonstrated how near-real-time analytics can improve outcomes through early alerts. Modern systems like Canvas Data Services or Arizona State University's learning analytics infrastructure are extending that capability, enabling instructional tools to adapt dynamically based on live student activity.

Ethical Governance and Transparent Stewardship

Yet even the most elegant and data-capable architecture is useless without trust. Universities are custodians of profoundly intimate data, the story of a person's learning life, and the rise of AI only deepens the moral stakes. When we layer in insights into well-being or soft skills, the sensitivity of data captured in the learner profile rises to the level of sacred trust. If students and faculty believe that data are being harvested rather than stewarded, never mind being sold or shared with others, the entire enterprise collapses. Responsible innovation requires strong data-governance frameworks, designated data stewards, and clear policies for accuracy, transparency, and ethical review, not to mention commercial level security and cybersecurity.

We must also be able to understand the why of the system's responses and interventions. Implementing audit trails—records showing how an algorithm or AI system arrived at a recommendation—enables institutional oversight and builds accountability. For example, if an adaptive learning system recommends a particular assignment sequence, instructors should be able to see the logic behind that choice. This transparency transforms AI from a black box into a trusted collaborator in the teaching process.

Ultimately, this ethical approach could distinguish universities from corporate AI providers. While large technology companies may build powerful learning tools, higher education institutions have an obligation to safeguard learner data. Students are more likely to trust systems built under the principles of academic stewardship than those governed by commercial interests. Yes, the public's trust in universities and higher education in general has diminished, but does anyone trust their personal data to the likes of Sam Altman, Mark Zuckerberg, or Elon Musk?

The Knowledge Graph: Replacing the Static Syllabus

To realize precision learning, the traditional, static course syllabus—a fixed document designed months in advance for an imagined average student—must give way to a dynamic, interconnected map of all knowledge and skills. This map, a knowledge graph, represents a body of knowledge as a network of entities (nodes) and relationships (edges). It is not just a list of topics, like a syllabus, but a structure that maps entities (concepts, skills, people, facts) and shows the relationships between them. Think of it like a subway map, not a table of contents. A traditional syllabus is a rigid list of stations you must visit in order. A knowledge graph is a subway map that shows every station, every transfer point, and every possible route. You can start at any station and see all the different paths, or prerequisite skills, required to get to a new destination.

The fixed syllabus represents a one-size-fits-all delivery model. The knowledge graph enables the *n*-of-1 precision learning model by serving three critical functions:

- **Dynamic customization:** When combined with a student's learner profile, AI can instantly traverse the graph. If a student already has mastery over the "cell structure" node, the graph tells the system to bypass that content and jump straight to a more advanced related node, like "meiosis." The AI generates the curriculum pathway in the moment, based on a real-time understanding of the student in that moment.

- **Real-time scaffolding:** If a student struggles with a concept, an AI learning agent uses the knowledge graph to immediately identify the fundamental prerequisite skills they might be missing. The computed curriculum then instantly generates a micro-module to address that specific gap before allowing the student to continue.

- **Measurable outcomes:** The graph defines exactly what constitutes mastery for every node and link. This allows the AI to track a learner's progress across a complex web of interconnected skills, moving beyond simple letter grades to demonstrate genuine skill development and transfer.

Switching from the syllabus to the knowledge graph has powerful implications. A knowledge graph does not just describe what students need to know; it reveals how knowledge itself is connected—the hidden threads that link an introductory writing class to a capstone project or a statistics course to a student's eventual work in public health. The traditional syllabus is dead in the world of AI-mediated instruction; the knowledge graph is the syllabus of the future.

At the program level, curriculum stops being a static list of courses and starts to look more like a network of competencies and pathways, a set of interlocking routes rather than a single prescribed road. When AI can read and reason across those connections, it can surface patterns we've never had the tools to see: the unexpected bridge between philosophy and coding, the way a class in environmental literature quietly predicts who will thrive in sustainability research, the skills that cluster among graduates who find meaningful work.

This is not about letting algorithms design degrees but rather giving faculty and curriculum designers new ways to perceive their own ecosystem, to see where learning outcomes overlap, where redundancies persist, and where interdisciplinary connections have gone unnoticed. The curriculum becomes less a catalog and more a living graph, dynamic, evidence-informed, and open to iteration. The faculty member is an architect of learning, partnering with AI to create knowledge graphs that go well beyond static syllabi.

Data + LLMs = Computed Curriculum

The convergence of robust learner profiles, data architectures capable of real-time intervention, the knowledge graph, and the generative power of advanced AI heralds the arrival of precision learning. LLMs can instantly ingest learner attributes—spanning background, intrinsic interests, prior coursework, and social context—to construct learning experiences that are profoundly relevant and engaging, including just-in-time content tailored to the learner and the moment. This is the core of the computed curriculum: generating content and activities dynamically, tailored for the unique individual.

Imagine the curriculum being spun up in the moment and delivered precisely when the learner needs it. In a foundational philosophy seminar, a student is not simply assigned a fixed reading on utilitarianism. Instead, the student might engage in a sophisticated Socratic dialogue with an AI agent that helps them explore the philosophical concepts through the lens of a current ethical dilemma in their stated field of interest (e.g., resource allocation in a modern city), drawing on tailored, just-in-time resources. The system identifies moments of confusion or mastery in real time, enabling the AI to adapt its direction and support to ensure sustained engagement and successful progression.

This principle scales to more sophisticated high-stakes learning interactions. In a master's program on public policy, the LLM can instantly generate realistic scenarios and dynamic dialogues. It weaves in elements from the learner's profile, perhaps their experience as a nonprofit manager or their expertise in urban planning, to create a hyper-customized policy case study or an executive role-play simulation. This not only dramatically boosts learner engagement but also enhances the crucial transfer of learning to complex, real-world professional contexts. Case studies, which are enormously time-consuming to produce, can now be spun up with lightning speed and customized for individual learners.

This kind of precise learning experience mirrors the gold standard of one-to-one interaction, whether that's historical apprenticeship or master-level tutoring. The effectiveness of these time-tested models rests on the mentor's ability to adapt their guidance based on the apprentice's lived context and demonstrated skill. An experienced mentor would not reteach fundamental budgeting to a finance major who had already mastered it; rather, they would elevate the instruction to advanced areas like specialty derivatives modeling. In those interactions, there is nuance, probing, adjusting up and down, and shifting modes to drive understanding. LLMs can facilitate this by processing a complex matrix of contextual variables to engage learners in natural language interactions that create a synthetic, yet authentic, connection between the knowledge base and the learner's current state. The key is the synergy between the precise individual learner profile and the LLM's capacity to weave that data into a live, fluid conversation.

LLMs + Learner Profile + Computed Curriculum = Precision Learning

The current moment marks a fundamental new phase beyond traditional personalized learning. The precision learning breakthrough requires two nonnegotiable components: modern data architectures that enable real-time intervention and the adaptive power of LLMs to deliver responsive, context-aware learning experiences.

The payoff is significant: students who feel genuinely seen, recognized, and supported demonstrate higher engagement and greater persistence and achieve superior outcomes. This value is centered not merely on efficiency, but on relevance, motivation, and equity. To realize this promise, institutions must invest in real-time data infrastructure and adopt pedagogical models that treat every learner as an individual. Done right, precision learning can finally deliver on the goal of education: providing each student the right support, at the right time, in the right way.

This is the new frontier: n-of-1 precision learning, where data unique to a single student is used to create a learning experience tailored exclusively to them, moving beyond broader demographic or performance patterns. For higher education, this shift presents formidable challenges that must be addressed proactively:

- **Evolving faculty roles and pedagogy:** Precision learning demands a fundamental rethinking of the faculty role, as will be explored in chapter six. Faculty time shifts from content delivery to becoming learning experience designers, knowledge graph architects, and high-touch human mentors. The AI handles basic scaffolding and content generation, freeing the professor to focus on high-stakes, uniquely human engagements such as ethical critiques, complex project oversight, career guidance, and personalized psychological support.

- **Balancing support and intentional challenge:** While precision learning aims to meet learners where they are, there is a critical tension between providing optimal support and unintentionally capping potential. If a system continuously shortens texts for a slow reader, it risks "over-adapting" and removing the productive struggle necessary for long-term growth. Finding zones of proximate development, precision learning must be designed to ensure interventions scaffold progress rather than protect learners from the experiences that foster development.

- **Rethinking evaluation, mastery, and evidence:** Dynamic content requires equally dynamic assessment. Fixed, point-in-time checkpoints (midterms, finals) and static competency maps will no longer suffice. Institutions need measures that capture nuanced longitudinal skill development, assessment models capable of tracking a learner's growth velocity and their ability to demonstrate the transfer of learning to novel contexts. Failure to update assessment risks applying outdated evaluation logics to a revolutionary new paradigm, with formative assessments giving way to ubiquitous, in the background, of assessment in real time.

- **The problem of shared experience and community:** When every student is on a unique pathway, we must intentionally curate the learning community, foster connection, and build human skills of empathy, collaboration, and navigating the "other." In a world increasingly fragmented, we must ensure that our AI models not only support better individual learning but also create space for more, not less, human connection. Universities and faculty members will need to think about the ways common experiences are developed and experienced, how students connect in curated classrooms within the curated community that universities must be.

- **Equity and ethical bias:** AI systems can inadvertently reinforce existing inequities through curatorial bias. Universities must rigorously guard against this, ensuring that the computed curriculum does not create biased learning pathways or unintentionally limit the long-term potential of specific demographic groups.

Managed wisely, these shifts can unlock unprecedented levels of individualized support, higher engagement, stronger outcomes, and more relevant educational experiences for every learner. In short, they can revolutionize learning.

The move to precision learning is no longer a theoretical ideal but a technical inevitability, driven by the convergence of real-time data architectures, the knowledge graph, and generative AI. The convergence of these elements offers higher education an opportunity to finally deliver on the decades-old promise of personalized learning and move into a new age of precision.

The Enormous Benefit: Equity and Personalized Support

The immediate and most significant benefit of precision learning is not efficiency, but equity. Precision learning democratizes the gold standard of education: the one-to-one tutoring and deep contextual awareness historically reserved for the privileged. For students who are underserved, underprepared, or dealing with complex academic and personal challenges, precision learning transforms the entire system's responsiveness. I previously named three teachers who transformed my life, but many students struggle to name even one. They have often had to navigate a system rather than a relationship, their struggles to learn often unnoticed or unaddressed except for poor end of the term grades.

The AI-driven curriculum removes unnecessary roadblocks. If a student is underprepared in a prerequisite skill, the system detects the gap in milliseconds and instantly generates the precise scaffolding required, ensuring they do not fail out of a program simply because of rusty knowledge. If a student is struggling with a personal challenge—such as fatigue, a dip in well-being metrics, or high cognitive load detected by system activity—the computed curriculum responds with customized supports like "mindfulness breaks" or a shift to lower-stress content modalities. Wellness agents can support student mental health, when carefully designed. Precision learning ensures the system sees and responds to the needs of the whole student, creating learning pathways that are truly optimized for their success.

When we first introduced College for America, SNHU's pioneering competency-based education program and the first direct-assessment program approved by the US Department of Education, I met one of our first students at a partner site. She was a single mother and her then-seven-year-old daughter suffered from a chronic illness. When the illness flared up, her daughter would miss elementary school for a week or more, and our student would miss classes at the local community college where she was enrolled in an associate degree program. As a result, her transcript was littered with failures and withdrawals. On paper, it looked as though she was not right for college. When we enrolled her in our self-paced program, she flourished. Whenever her daughter was ill and needed extra attention, our student simply "hit the pause button," as she described it. That little bit of personalization and flexibility enabled her to take care of her daughter *and* complete her degree. It turned out she was smart, hardworking, and capable. Instead of not being right for college, college was not right for her.

Imagine what might be possible for students not served by a one-size-fits-all education model that forces every student to learn on the institution's terms and timetable, as opposed to a model that meets students wherever they begin, conforms around them, and provides them just the right experience to maximize their chance of success. A system that routinely fails 45 percent of those who start and has produced more than forty million learners with some credits, no degree, and typically debt is ready for dramatic change.

The Enduring and Elevated Role of the University

That change, made possible by the technological revolution underway, does not diminish the university's mission. It elevates it. As AI handles the labor of content generation, basic scaffolding, and knowledge transfer, faculty and the institution are free to focus on the high-stakes human work that is more important than ever:

- **Knowledge engineering and curation:** Faculty become knowledge-graph architects, responsible for structuring the curriculum, defining the connections between concepts, and certifying the quality and ethical neutrality of the knowledge base. They stay on top of a fast-changing world of work, ensure that programs align with workforce needs, and become expert in the tools of knowledge mapping. They curate their classroom experiences to enrich and develop human skills.

- **Intentional learning communities:** When the pathway to mastery is unique for every student, creating intentional, diverse learning communities becomes a crucial university function. The focus shifts to facilitating complex collaboration, developing empathy, and navigating the "other" in shared high-touch, human-led projects. When programs are so highly individualized, the need for common and shared experiences becomes critical, especially when grounded in values. We do not want universities to exacerbate the atomization and alienation already afflicting society. Universities can become part of societal healing and nurturing human skills through carefully curated communities.

- **The human dimension and mentorship:** To that last point, the faculty's role is transformed into that of the human mentor. Freed from routine lectures, they can dedicate their time to ethical critiques, complex project oversight, career guidance, enrichment, and personalized psychological and human support—the very interactions that define a truly transformative education.

- **Certifying learning:** The university's role as the certifying body for competence and skill transfer remains indispensable. Dynamic assessment embedded within the knowledge graph enables the university to provide richer, longitudinal evidence of genuine mastery, ensuring the credential retains its value in an automated world.

However, achieving precision learning demands an enormous, often painful institutional transformation. It requires a radical rethinking of faculty roles and program design and delivery, along with a massive upgrade to institutional digital infrastructure.

The most critical challenge is that universities must, in many ways, begin to think of themselves as data organizations. They lag far behind the commercial sector in real-time architecture, data governance, and data literacy. There are immense ethical and logistical complexities of moving from fragmented data silos and old batch processing to event-driven architectures that enable real-time intervention. This modernization is not optional. In fact, it is the infrastructure necessary to power the ethical, equitable, and effective use of AI. It is probably the most urgent need on most campuses at this moment. It is all about the data.

The choice is stark. Universities can remain bound by infrastructures built for another era, or they can embrace systems that enable real-time care and insight. Their unique mission, rooted in trust and human development, is their greatest asset. By combining that mission with a commitment to modern data stewardship, universities can offer an education that is perfectly precise, deeply human, and fundamentally equitable, solidifying their role as indispensable agents of transformation in the Age of AI. Computed curriculum, precision learning, and carefully curated values-based university communities can make possible for every learner that which was promised only for the privileged and the few, an educational model for the Age of AI.

6

What Is a University to Do?

"What should we be doing?" and "Shouldn't we be doing more?" These are the two questions I keep hearing from presidents and boards of trustees in what have become almost weekly calls. Artificial intelligence (AI) is everywhere. Students are using it in ways both sanctioned and not. Faculty are experimenting and often complaining about it or demanding outright bans. Institutions and systems are announcing enterprise licenses with one large language model provider or another, often with more fanfare than planning for what they will do with the technology. The workforce, already affected by economic uncertainty and political volatility, is being reshaped by AI, affecting the upstream work of universities to produce workers and create opportunity and increasingly leaving it on shaky ground. "What to do?" indeed.

There is some irony that the very institutions devoted to understanding and explaining the world have been caught somewhat unprepared for the most transformative technology since the printing press. Across industries, AI is refashioning the fabric of work, creativity, and human interaction. Banks use it to assess risk, hospitals to diagnose, law firms to review contracts, and artists to generate images once thought to require the ineffable spark of imagination. Our universities will be no exception. The question is not if higher education will change in response to AI, but how, and whether it can do so in ways that reinforce its deepest commitments rather than betray them.

The paradigmatic changes argued for in the preceding chapters will be enormously hard to realize. Rebalancing knowledge and human development, shifting focus from product to process, reimagining faculty as curators of community, transitioning to n-of-1 learning and ending one-size-fits-all program design—these are huge. Those bigger questions beg the immediate chaos of making technology choices, training and supporting stakeholders, and rethinking disciplines so students are prepared and competitive in a tough labor market. Simply making sense of it all. There will be enormous pushback and resistance.

My first bit of advice when those anxious calls come in is, "Take a breath." We have some time, even as the steady drumbeat of announcements and new releases from the large AI companies make the velocity of change feel as though they are happening at warp speed.

Why say, "We have time?" For the following reasons:

- New releases do not mean good releases. In the rush to outdo each other, to keep valuations high, and to get market share, the big AI companies often get it wrong. OpenAI, arguably among the most reckless of the providers, released GPT-4o in April 2025, and in its attempt to be helpful, was so sycophantic as to be deemed dangerous. Carnegie Mellon University's Maarten Sap cited the way 4o could reinforce biases and even encourage risky or dangerous behavior.[1] One colleague of mine asked it, "Am I the best person who ever lived?" It replied with something like, "While that is a complicated question to answer, when considered broadly, it is fair to say you are one of the best people alive." Great, just what a world of Instagram narcissists needs: reinforcement. OpenAI quickly rolled back that release. There also was Google's Gemini struggle with inaccurate and in some cases zany image generation; Perplexity's plagiarism and made-up facts in 2023 and 2024; and the early and infamous example of Microsoft's 2016 chatbot, Tay, which mimicked its users' racist and misogynistic sentiments within twenty-four hours of its release.[2]

- There has been the inevitable pushback and backlash. As mentioned previously the 2023 writer's strike in Hollywood might be seen as the first labor action against AI, with its threat to the creative professions increasingly apparent and resulting in several key agreements to control its use in the final contract agreement. Regulatory actions will slow and shape adoption of AI in many industries, and while the federal government is holding to a laissez-faire approach to any regulation, obsessed with the race for AI dominance with China, the US Food and Drug Administration and state regulatory bodies are stepping in. For example, under its Unfair or Deceptive Acts or Practices code, California has laws demanding transparency in the use of AI for decision-making in lending and employment. California, which provides the most active regulation of states in this space, also prohibits AI systems from implying that they have licensed medical oversight when they do not. In higher education, the European Union's AI Act identifies "high-risk" areas in education, such as evaluating learning outcomes, proctoring, and admissions processes, and requires robust "human in the loop" oversight. These could very well become a model in US higher education.

- This may surprise no one, but higher education changes slowly. Because it is a regulated industry, it has never experienced the disruption that fundamentally changed media, financial services, or entertainment. While that glacial pace puts it at grave risk in terms of relevancy, value, efficiency, and cost, it has buffered it from the cataclysmic change experienced in those other cited industries.

So we have time. But not endless amounts of it given the already precarious state of American higher education. Resilient as the sector has been, it feels like we are reaching a tipping point in which some combination of demographics, broken business models, loss of public support, and poor outcomes marks the end of an era in higher education. Historically, when higher education was found wanting, unsuited for the demands of the time, it went through enormous change. When Abraham Lincoln, amid national collapse and civil war, signed the Morrill Land-Grant Act in 1862, he initiated one of the most profound yet unexamined overhauls of American higher education. This was not a modest amendment, but a wholesale federally-subsidized expansion declaring that the classical European curriculum—all Greek and theology—was no longer fit for a modern industrial and agricultural nation. By dedicating vast swathes of federal land to fund public colleges, the act built an entirely new university system from the ground up, forcing institutions to teach the "mechanical arts" and agriculture. In that respect, it democratized college and positioned the American university for the first time as an engine of economic utility rather than purely philosophical contemplation.

Another example of tectonic change in higher education came with the Servicemen's Readjustment Act of 1944, known as the GI Bill. Where Morrill created the public framework and the *why* of modern higher education, the GI Bill provided the overwhelming *who*. It was an unprecedented act of federal investment in human capital that did more than merely pay tuition. It transformed the social hierarchy, injecting millions of veterans, many of them first-generation students, into a system that was scarcely prepared for the sheer scale of the influx. Suddenly, the university became the primary engine for the American middle class, permanently shifting the perception of a college degree from a cultural luxury to a necessary economic credential. It also forced institutions to rapidly expand their infrastructure and faculty into the sprawling, comprehensive credentialing machines we recognize and increasingly criticize today.

While perhaps not as tectonic and certainly not as top down, the cultural reckoning of the 1960s, driven by student demands and politicized campus debates, led to massive change. We saw the rigid paternalism of *in loco parentis*, where the university acted as a surrogate parent, unceremoniously dismantled, and the great required core curricula shattered in a bid for "relevance." Colleges conceded the power of choice to students, leading to the proliferation of free electives, the birth of interdisciplinary studies, and the establishment of new departments like Black Studies and Women's Studies, which forced the academic canon to acknowledge its omissions. This was not a change legislated from above, but a cultural concession won on the quad, dramatically shifting the university's relationship with its students from one of *dictation* to one of *negotiation*. The result was an educational structure that was dramatically liberalized yet, depending on one's perspective, perpetually in search of its lost rigor.

We are too much in the middle of it today to know if we are in another one of those transformational inflection points for higher education, but four dynamics suggest we are:

- Many find higher education wanting for the time in which we find ourselves, reflected in a dramatic loss of consumer confidence, questions of its value and return on investment, a new openness to alternatives, and debates about its relevancy.

- The activist engagement of the federal government is seeking to reshape the sector, which was pronounced during the Obama administration but more like a full-frontal assault in the Trump administration, with a strong desire to roll back the political liberalizing of the campus that started in the 1960s and to push for more focus on workforce and outcomes.

- There is a countervailing pressure from students and some faculty making demands of the university around fraught political issues. That is no better illustrated than the Palestine-Israel protests that roiled campuses in 2024, leading to mass arrests, violent clashes, congressional hearings, presidents losing their jobs, and new controls on speech and action on campuses. Those protests have become muted, but the response to funding cutbacks and the "compact" put forward by the Trump administration have taken center stage on many campuses.

- AI has arrived on our campuses, with one survey suggesting that somewhere about 90 percent of university students are active users.[3] That number is echoed in global surveys, with 86 percent of students in sixteen countries reporting regular use.[4]

While there may still be time, AI is here and some aspects of it deserve immediate attention. If we are indeed in one of those transformational moments, AI is a forcing function, as argued, and a critical disruptive technology that we can harness and put to work as we reimagine higher education for the Age of AI.

To move from rhetoric to readiness, institutions will need a deliberate phased approach, one that begins with clear ethical guidance, progresses through curricular and infrastructural realignment, extends into a forward-looking theory of work, and culminates in a deep reimagining of the university's very purpose in an emerging new world order created by AI.

Phase One A: Institutional Guidance and the Ethics of Use

The first order of business is stakeholder clarity. Universities need to establish shared norms for how faculty, staff, and students should and should not use AI. Without them, institutions risk lurching between permissiveness and panic, celebrating innovation one day and policing it the next. I often start a conversation with a university president by asking about campuswide guidance on the uses of AI and how it is working. With distressing frequency, I hear, "We leave it to the faculty members and department heads." While there are faculty who want to make those decisions based on principles of academic freedom and autonomy, and others who hate the idea of AI and how they see it being used and are happy to mostly ban it, the great majority are not equipped to make those calls.

Though most students are using AI, most also feel a lack of support from their institutions and want more training, support, and guidance. Many have a nuanced sense of what should be in and out of bounds for their use of AI in coursework, arguing, for example, that they should be able to use it in research, brainstorming, and notetaking. However, most students see using it to write as cheating, and few resist requirements to reveal when and how AI was used in their work.[5]

Understanding that the fast-changing nature of the technology and the evolving nature of how we as humans are using it will require frequent rethinking and revisions, and we are seeing thoughtful attempts at good guidance taking place, like the "Student Guide to Artificial Intelligence."[6] Individual institutions also are stepping up. At Stanford, the Office of the Provost issued early guidance acknowledging that generative AI could be a powerful tool for creativity and research but also warned that the university considers AI-generated work submitted without acknowledgment to be plagiarism. Harvard, by contrast, framed its guidance as a living document, evolving alongside the technology. Others, like the University of Sydney, have integrated AI literacy into academic integrity policies, explicitly requiring students to disclose when they have used AI in assignments.

Such institutional coherence does more than prevent misconduct; it models the ethical reasoning universities exist to teach. Students should learn not merely *how* to use AI tools, but *when* and *why*. A philosophy course might discuss the ontology of machine "understanding," a computer science seminar might explore the biases embedded in training data, a journalism class might debate whether using AI to draft a paragraph constitutes a breach of ethics.

Thus, training support and equitable access must accompany sensible policy. It is no longer sufficient to let students experiment ad hoc with whatever AI tools they discover online. Letting individual administrative units select and deploy AI solutions on their own is a recipe for eventual chaos, not to mention data breaches. Universities should identify and provide a suite of approved secure AI applications, ideally ones with educational licenses that respect privacy and intellectual property. Faculty, too, need structured opportunities to learn how AI can augment their teaching and research. The University of Michigan launched the Generative AI Lab where faculty and graduate students test tools like ChatGPT, Midjourney, and GitHub Copilot in pedagogical contexts.

In short, Phase One A is about building a foundation of *trust and literacy*, the shared ethical infrastructure upon which everything else rests.

Phase One B: Curricular Alignment, Data Readiness, and Administrative Transformation

Once universities have established ethical guardrails, they must turn inward to examine their academic structures and institutional machinery. This phase concerns the pragmatic work of aligning education with an AI-shaped economy, strengthening data infrastructure, and modernizing administrative operations.

Reimagining Academic Programs

Every department should begin by asking two deceptively simple questions: What is the world our graduates are entering? What will it demand of them? This requires more than a one-time curricular update. It calls for a systematic audit of programs: identifying which disciplines are being reshaped by AI automation, which skills will become newly essential, and which human capacities remain uniquely irreplaceable. If I were a president, provost, or dean today, I would want reassurance that my faculty and department chairs are urgently reviewing every course and program for the AI tools that graduates will be expected to use, for the ways the discipline is being changed, and for those shifts to be reflected in the curriculum. This cannot be completed at the typical measured pace of most curriculum review and revision. AI is changing work too rapidly, and recent university graduates are facing a tough job market: 40 percent of US business leaders say "junior/entry level roles had been reduced or cut due to AI."[7] In China, 45 percent of business leaders say "they expect *their* [emphasis mine] current job/role to no longer exist by 2030 because of AI."[8] Our academic departments can no longer prepare graduates for the entry-level roles of yesterday. They must quickly recalibrate, find ways to help them enter the workforce at the higher levels not previously possible, and ensure that they are equipped to work with AI.

As argued previously, preparing graduates is no longer a question of knowledge and using the right workplace tools. It requires a much greater emphasis on human skills. In nursing, for instance, AI assists in diagnostics and patient monitoring, but empathy, ethical judgment, and embodied care remain irreducibly human. In law, machine learning can draft contracts and summarize cases, yet the ability to interpret ambiguity and argue nuance will distinguish tomorrow's practitioners. And in the humanities, AI's facility with text generation only heightens the need for people who can interpret, critique, and imbue meaning into words.

If those examples point us to what might be broadly called critical thinking skills, we must also be more intentional in developing emotional intelligence and interpersonal skills, as noted in previous chapters. This is true even as students arrive with greater deficits in their human skills, reflecting some combination of COVID-19-driven social isolation and the impacts of lives lived on social media platforms. Professionals will still need foundational knowledge, theory, and understanding, alongside mastery of demonstrable workplace skills, and those will also need to be integrated with the human skills that come with a more balanced mix of the epistemological with the ontological.

Data Infrastructure: The Nervous System of the Modern University

Yet such curricular reform will falter without the underlying capacity to understand students and outcomes in real time. As previously described, AI thrives on data, and most universities remain hamstrung by fractured systems—siloed student information, inaccessible learning analytics, and outdated data governance. Consider the student life cycle: an applicant's detailed engagement history (like attending five virtual events or reading specific major pages), stored in the admissions customer relationship management system, becomes completely invisible once they enroll. This crucial behavioral data never transfers to the advising office's student information system, leaving advisors blind to the student's initial interests or potential risk factors when crafting an academic plan. This fragmentation forces staff to waste time manually reconciling reports or, worse, leads to missed opportunities for personalized support, ultimately detracting from the university's core mission of student success.

This challenge extends deep into operational and research activities, preventing the kind of data-driven efficiency modern institutions require. For example, in financial management, the procurement system tracks real-time spending, but the budgeting system used by department chairs tracks only original allocations. Because these systems do not talk to each other, department leaders often must wait for weekly or monthly exports to find their true available balance, hindering timely decision-making. Similarly, the facilities management system that tracks energy consumption often operates independently from the room-scheduling system, meaning the heating and cooling system might run full blast in an empty lecture hall simply because no one told it that the class was canceled. These pervasive noncommunicating systems erode institutional efficiency, increase operating costs, and prevent the university from having a unified three-hundred-sixty-degree view of its most vital assets: people and resources.

There are some excellent new AI point solutions being developed for higher education. But the kind of enterprise-wide and integrated AI models that could transform a university's operations and create far better learning? Results for students remain elusive as long as data sits in silos like the ones just described. If AI is a Ferrari, data is its fuel: data readiness is the most significant barrier to effective AI deployment in higher education.[9] The challenge is not simply volume but coherence: bringing together learning management data, advising records, enrollment patterns, and alumni outcomes into interoperable systems. Georgia State University's predictive analytics system and Southern New Hampshire University's Unify customer relationship manager demonstrate how unified data architectures can inform intervention, retention, and personalized learning.

To realize the full potential of AI, institutions will need to adopt *data fabrics*, ecosystems in which secure, standardized application programming interfaces allow different platforms to exchange information seamlessly. This enables what technologists call *event-driven architecture*, where every student action (a quiz attempt, a discussion post, a missed login) generates a data event that can trigger timely support. Such responsiveness transforms data from static archives into living systems of care. Long, well-established market leaders like Ellucian are racing to integrate AI and the various solutions they offer with the student information system that is the central nervous system of any university. Meanwhile, startup companies like K16 Solutions (full disclosure: I serve on its board) are working to pool, clean, and make universally accessible institutional data in next generation data lakes.

The modern university will in many ways be a data organization built to deliver learning, not a learning organization that happens to have a lot of disparate data. Data governance must accompany this technical evolution. Universities should appoint data stewards, define clear standards for accuracy and privacy, and maintain audit trails for AI decisions. The goal is to ensure that AI systems operate transparently and ethically—not as opaque arbiters, but as accountable and responsible collaborators. And in what is admittedly a hard sell, I tell presidents their data infrastructure should be a priority and not simply treated as a part of institutional infrastructure whose management is beyond their pay grade.

Rethinking academic programs may be urgent but will take longer than anyone would want because that is simply the nature of academic evolution. However, once the university's data infrastructure is well engineered, the administrative backbone of the universities can be transformed. Much of the early productivity gains from AI will emerge not in classrooms but in offices like human resources, communications, marketing, admissions, and facilities. At Purdue University, AI scheduling software optimizes classroom use and maintenance cycles. At the University of California San Diego, admissions officers are piloting AI tools to triage applications, freeing staff to focus on holistic review. The University of Michigan's finance office has implemented a conversational AI that allows staff to query budget data in natural language: "Show me travel expenditures by department for 2024."

These efficiencies are not trivial. A university that reduces administrative overhead by even 5 percent could redirect millions toward research and student support at a time when fiscal pressures are at a crisis level. Moreover, such systems generate new data streams that can improve decision-making. AI can redefine the administration of the new university.

Phase Two: Developing a Theory of the Future of Work

The work of Phase One is about foundation building: setting good policy, putting in place supports, getting data infrastructure right, then making sure academic programs are revised to ensure graduates are ready for today's workforce needs. But what of tomorrow? Universities must have a theory about the future of work against which to map their curricular offerings. The World Economic Forum estimates that 44 percent of workers' skills will be disrupted within five years.[10] McKinsey & Company projects that in 2030, nearly 30 percent of current tasks in the US economy could be automated, though many jobs will also be redefined or created.[11] Previous technological revolutions resulted in the creation of whole new job categories. In this technological revolution, it is unclear whether many if not most of those newly created jobs can also be done by AI. We know we will face at least a massive reskilling challenge if McKinsey is right. I am in the more radical camp that argues that AI will do to white-collar jobs what automation did to blue-collar jobs, that we will see massive job displacement. This is a prediction dating to a 2013 Oxford study, well before the advent of generative AI, that even then estimated that 47 percent of all jobs in the United States could be automated or computerized by the early 2030s.[12] The point is that universities must develop and keep updating some theory of the future of work that can then guide the management of their academic program offerings.

The importance of those models is critical: if universities are in the business of preparing people for the future, they must have a coherent theory of what that future entails so they can have the right programs to serve the nation's workforce needs, and the right programs to create opportunity for their students.

This requires an analysis of the following issues:

1. **Jobs likely to disappear or be fully automated:** Routine accounting, data entry, transcription, and certain forms of legal or medical review seem almost a certainty. We have seen computer science and coding jobs implode in 2024 and 2025, so I would not bet on the long-term prospects of accounting or other information processing jobs.

2. **Jobs likely to be redesigned:** These professions will blend human oversight with machine efficiency: teachers, nurses, engineers, analysts. It will be the rare job that does not include AI partnership, thus changing the work the humans do in those partnerships quite different and, as has been argued, far more human focused.

3. **Jobs likely to be created:** AI ethics officers, data stewards, human-AI interaction designers, prompt engineers, and cognitive system trainers are all likely jobs, especially if we hope to see a responsible and transparent deployment of AI in our society (a touch-and-go question at the moment, I fear).

Some universities are already anticipating this reconfiguration. Northeastern University's humanics curriculum emphasizes the integration of data literacy, technological fluency, and human-centered design as the foundation for all programs. Similarly, the University of Sydney's Future of Work Institute conducts ongoing labor market analysis to inform course development.

A university that takes this phase seriously might develop its own Work Futures Council, drawing from economics, sociology, and computer science to model scenarios for the next decade, monitoring and distilling the growing flood of workforce reports and predictions. These insights should guide program creation and closure alike. Just as the liberal arts once organized on the trivium and quadrivium, the university of the AI era may organize on a new triad: *data*, *design*, and *ethics*.

But this phase also demands a deeper intellectual humility. The university must resist the temptation to treat employability as its sole metric of success. It has the greatest urgency, and AI is already changing the texture of work, but it also raises profound questions about meaning, identity, and purpose—questions that only humanistic education can answer. Which brings us to Phase Three.

Phase Three: Reinventing the University

The final phase is less a plan than a provocation. It asks what a university is *for* in a world where knowledge can be generated, translated, and transmitted by machines. If AI can summarize any text, generate any image, and simulate any experiment, is what remains the unique province of the university?

The answer, paradoxically, may lie in returning to its oldest purpose: cultivating wisdom. Knowledge is now cheap, but discernment is priceless. The university must become a place where machines handle information and humans handle meaning. I did not expect to turn to Cardinal John Henry Newman's 1853 *The Idea of the University* for inspiration on education in the Age of AI, an age he could not begin to imagine, but his work offers a brilliant lens through which to view the modern challenge, which comes as a bit of a surprise given my lack of interest in it when it was assigned to me as a student. But his treatise is surprisingly relevant, as his core philosophy perfectly validates the call for wisdom over what he called "cheap knowledge." He defined the university's purpose as the cultivation of "intellectual culture," which he saw as a mental development superior to mere professional training or the absorption of facts. For Newman, the true output of a university was not a technician equipped with a list of facts, but a person possessing "a comprehensive view of the scope and bearings, the use and abuse of every subject," ready to apply reason and judgment to any field.[13] Because AI handles the "scope and bearings" (information) instantly, the university must focus on teaching the judgment—the ethical, human ability to distinguish truth from plausible imitation—that only the philosophical habit of mind can acquire.

Newman's secondary but equally vital argument was that the university is fundamentally a social space, a community where scholars live and interact. He believed that the education gained from "the habitual living with men [it was really just men back in those days] and acting among them" was indispensable.[14] The idea of the university as an intentional community and the teacher as a curator of *human* interaction in the classroom echoes Newman's assertion that "humans handle meaning." While the machine can automate analysis, it cannot replicate the complex, ethical, and collaborative negotiations required to make meaning and application of that analysis in a human context. That human context often includes the irrational, the passions, and all the messiness of what it means to be human.

We have many examples of institutions that come closer to Newman's ideal because they boldly assert their values, their existential stance. While they still train people for the workforce, and very specific kinds of work in the case of military academies, values-based institutions also embrace the questions of what it means to be a good human, at least as understood through their lens. At Jesuit universities (which emphasizes *cura personalis*, or care for the whole person) or Brigham Young University or Yeshiva University, the curriculum is intentionally structured on a unified moral and theological center. In historically Black colleges and universities, the mission is often inseparable from the shared history of resilience and community uplift. A Morehouse Man knows what it means to be a graduate of that institution. Similarly, at military academies, every course—from calculus to classics—is taught within an explicit, nonnegotiable framework of duty, honor, and country. This ensures that students are not just absorbing cheap knowledge, to use Newman's phrase, but are being trained rigorously in discernment, forcing them to integrate technical skills with high-stakes ethical judgment. By prioritizing the formation of character and leadership, the core of human wisdom, these schools maintain the ideal of the university as a unified cultural engine, focused on who the student *becomes*, rather than merely what they *know*.

The reinvention required of universities may be philosophical rather than structural: the university as the site not just of skill acquisition but of moral orientation, a place that teaches what it means to be human alongside intelligent machines. In this reimagined model, universities function less as dispensers of content and more as *curators of transformation*. The lecture may give way to the learning studio, the degree to the lifelong learning graph. Students might assemble modular credentials across institutions and industries, guided by AI mentors but anchored in human mentorship and engagement. Assessment could evolve from standardized tests to portfolios that reflect a learner's journey.

If the passage of the Morrill Land-Grant Act in 1862 was intended to address the needs of an increasingly industrial economy, the challenge to rethink the purpose of a university education may bring us full circle to that European classical education the Morrill Act sought to replace, or at least balance. One might even argue that if the university must shift its focus to moral orientation and teaching "what it means to be human," it must in some manner make the humanities' purpose the *central mission* of the entire institution. It changes the value proposition from what you know (a machine-dominated question) to how you judge and why you act (a uniquely human endeavor). However, before my old colleagues nominate me for sainthood (I was an English professor and chair of a humanities department early in my career), the humanities as structured and taught in the old-world order will not suffice.

Just as Carlota Perez argues that the state must eventually answer the question, "What does a good life look like for its citizens?" but not the state as it existed before, we must ask "What does a university do for its students?" but not the university or the humanities as we knew them before.[15] That question should be treated as an invitation, not a threat. The questions are to be relished:

- How might we rethink the old disciplinary boundaries (and boundaries altogether)?

- How might we better ground the humanities in the real world so they feel more relevant, more exciting, and more transformative?

- How do we make the humanities as much about individual transformation and growth as about knowing the canonical works, and maybe even making them fun along the way?

- How do we teach the humanities in a way that does not suffocate the thrill, emotion, and love that often draws students to their study with the technical and structured study of those disciplines?

I focus on the humanities here, but the arts are just as important. Whether music, sculpture, dance, painting, or creative writing, the arts remain essential for training creative intelligence and embodied experience, skills of innovation and human complexity that machines cannot replicate. Yes, it may be true that some combination of AI and robots can someday perform *Swan Lake*, but the novelty of that will last five minutes and will have none of the suspenseful beauty that makes us hold our breath when a ballerina executes a fouetté during that ballet's climax. AI might readily imitate the scene in Tolstoy's *Anna Karenina* when Kitty and the fumbling Levin instantly and intuitively realize their love for one another in ways beyond language, but I doubt its ability to ever create such an original scene and moment.

The Age of AI might allow us to enter a new Golden Age in which human-centered work and systems of care replace Information Age jobs, making communities and society better, while giving the people who do the work a sense of fulfillment and meaning that far too many lack today. Similarly, that Golden Age could save the university, reawakening its oldest reason for being. When information is infinite, the task of education might once again become to cultivate judgment, empathy, and purpose. The future of the university does not have to be post-human; it can become profoundly human.

To strategize for AI is, ultimately, to strategize for humanity. The university's task is not to compete with machines, but to *partner* with them in service of human flourishing, to rethink the world as it might be, not to cling to a world that feels increasingly broken. That partnership will demand ethical clarity, infrastructural rigor, curricular courage, and philosophical depth.

In the near term, this means the following:

- **Phase One A:** Clear institutional guidance and ethical frameworks, training and support, and urgent reworking of majors to make sure students can compete for jobs

- **Phase One B:** Robust data systems, administrative AI, and program realignment

- **Phase Two:** A rigorous theory of work and skill transformation, with a corresponding strategy for rethinking the academic program portfolio
- **Phase Three:** The reinvention of the university as a moral and intellectual commons for the Age of Intelligence

These phases should not be strictly sequential; much of the work can happen in parallel.

That said, there is growing impatience with higher education, and a phased approach such as the one recommended here will feel to many like inadequate incrementalism when AI is racing along so quickly. That's a fair critique, but I have twenty-eight years of being a president at two institutions and ample scar tissue from trying to push faster than institutional capacity can withstand. While the level of capacity for speedy change varies greatly across institutions, no one of them is winning a prize for agility any time soon. As such, these phases do not represent a leisurely timeline but the irreducible complexity of institutional change. Moving to Phase Three requires the ethical clarity (Phase One A) and infrastructural rigor (Phase One B) to support it. Trying to implement the "moral and intellectual commons" without the ethical and data foundations to support that work is not speed, but rather a kind of recklessness. We must move at the speed of a culture's ability to absorb change, not merely the speed of technology. Each phase is designed to be undertaken with the same urgency that chapter one suggests, but that will look very different from one institution to another. However, if universities can move through these phases with integrity and imagination, they may find that AI is not an existential threat but an invitation, a call to remember why learning exists in the first place.

The task before higher education is to ensure that explanation—and meaning—remain our province. In the Age of AI, the university's role will be what it has always been at its best: to cultivate minds capable not only of understanding the world but of remaking it wisely. To the skeptic who asks, "How does a better-run university actually fix a broken world?" the answer lies in the graduates we produce. That's always been the answer. By cultivating minds capable of remaking the world, wisely, we are not merely improving graduation rates but empowering the workforce and the body politic with ethical clarity, discernment, and the skills of care. Reinventing our institutions of higher education is therefore the greatest leverage we have to ensure the Age of AI leads to a more just and humane society, a human-centered society, not just a more efficient one that contributes to further wealth inequity, civil fracture, and a loss of empathy and community. For all our faults, and higher education has many, I would rather see our future shaped by our universities than by the technology companies vying for that power.

The Role of the Faculty in the University of the Future

My first experience teaching was when just a week before the start of a new semester I was given a teaching assistantship at Boston College, where I was a newly enrolled graduate student in the master's in English program. I was handed the pile of books that all teaching assistants (TAs) were supposed to use when teaching the first-year writing course required of all undergraduates. They included James Boswell's *Life of Johnson*, a collection of William Butler Yeats's poetry, William Shakespeare's *The Tempest*, and Virgina Wolfe's *A Room of One's Own*—all wonderful works, but maybe not the best choices for engaging students fresh out of high school. As immodest as it sounds, students loved my class, telling me so in person, echoing the sentiment in course evaluations, reminding me of it when I would run into them years later.

Why? It was not the material, as much as how I worked to make it come alive and seem relevant. It may have helped that I was not much older than my students, sported long hair and an earring, and shared many of the same pop culture reference points. But I think the real reason was how I responded to them: I loved my students. I got to know them, gave them lots of my time, listened seriously to concerns wholly unrelated to class. I brought them to a live performance of *The Tempest*, the $4.99 (hey, it was a lot of money back then) all-you-can-eat pasta night at a local restaurant, and the occasional lunch in the faculty dining room when someone needed to talk. As a college president three decades later, I was meeting with a peer from another institution when his general counsel stepped in to reintroduce himself to me. He was in that class, one of those who needed counsel when he found himself unhappy with his father's choice of major for him, and we talked about it over lunch in that wood-paneled dining room. He said, "That conversation changed my life. When I went home that Christmas break, I told my father I would never be a doctor and I was switching majors. He didn't like it, but he got over it and I've loved my career as a lawyer. I never had a chance to thank you for what you did for me." I do not think I ever gave him actual advice. I just asked him questions, and he found the right answers himself. I made him feel like he mattered.

Mattering is a term at the heart of social psychologist Greg Elliott's work. For Elliott, mattering is a core psychological concept defined as "the sense that one is a significant part of the lives of other people, institutions, one's community, or society as a whole."[1] He sees mattering as our core existential need, helping address the fundamental human questions of "Who am I?" and "Where do I fit in?" When we feel we do not matter, humans will often find ways, even deeply destructive ways, to assert themselves. That insight echoes research into attachment that shows how we seek connection and recognition with biological urgency. That is one of the core messages of Frantz Fanon's seminal *The Wretched of the Earth*, which examines the dehumanization of the oppressed and how they must inevitably violently rebel. It helps explain the murderous actions of the protagonists in Richard Wright's *Native Son* and Albert Camus's *The Stranger*, in which the main characters hope people will attend their execution and howl for their demise because at least they will know they *mattered*.

In a positive framing, mattering even to just one person can transform a life. Donna Beegle, who writes about people who somehow escape extreme poverty, says they rarely cite a program, policy, or organization, but rather a person, someone who made them feel like they mattered, gave them their support, and helped them dream bigger dreams for themselves.[2] Her point is borne out in research on developmental resilience by scholars such as Emmy Werner and Michael Rutter, who found repeatedly that the presence of *just one* caring adult often predicts whether a young person can thrive despite hardship.[3]

When I wrote my 2022 book *Broken: How Our Systems of Care Are Failing Us and How We Can Fix Them*, I asked Georgetown's Matt Biel, a specialist in child and adolescent psychology, why some children survive and even thrive in terrible situations when so many of their peers do not. He spoke of the importance of having some glimpse of a "normal" life, some picture of what better could look like, some passion that captivates them, whether it be music or basketball or astronomy. Most important, he said, they need *one person* who believes in them, who gives them time, who helps them lift themselves up. It's why I titled the first chapter of that book "Mattering." What I learned from Elliott, Biel, and other psychologists I interviewed was that one cannot truly transform a life if that person does not feel like they matter to you. I think that explains my success as a teacher. My students mattered to me and they knew it. Years later, as a leader, I worked hard to help every employee, from housecleaner to executive vice president, see how they mattered in the work we did together to educate students.[4]

I made this point in an interview about *Broken*, and the journalist, who focused on secondary education, said to me, "If I tell the teachers I've met in under-resourced schools, with too many kids in their classes and too many other things to do, from lunch duty to lesson plans to committees, that they have to spend more time knowing their students, as you suggest, they would roll their eyes and tell you that is impossible." I agree. Our systems of care, from education to healthcare to mental health to geriatrics, are designed to maximize efficiency and minimize time for the human relationships that should be at the heart of the work. A doctor who spends too much time getting to know each patient, an aide who chats for too long with a lonely elderly resident of a care home, and a faculty member who focuses too much on students and not enough on scholarship, publishing, and committees are all at risk, all disincentivized by the systems in which they work.

As I argued in that previous book, the key to making these systems *human-centered* again is to hold as sacred and most important the human relationship at the core of the work, to recognize the work of mattering as the heart of these profession, and to use technology thoughtfully to automate and support everything that does not require a human soul.

AI will now enable us to do just that in education. As seen in previous chapters, we have reached a point when the work of knowledge transfer is better done by AI. We are seeing advances in AI that can provide precision learning fine-tuned to the needs of each student, available twenty-four/seven, nonjudgmentally. These systems can draw on their subtly humanlike qualities to help students feel that they matter, while also supporting their well-being, strengthening their soft skills, and assessing their work in real time as part of their learning. Think of AI as a *genius TA* that will take over much of basic knowledge transfer and holistic support, unlocking learning when students get stuck, and helping assess students. Such a genius TA would readily provide dashboards to teachers—updating student progress, flagging those who are struggling, offering suggestions for what students might need from the faculty member.

What then of the classroom? The teacher? Freeing faculty to focus on building genuine relationships with students, using the classroom to foster human skills, and curating community may be the great gift of AI. But is it a gift faculty members want? It requires a profound reimagining of faculty roles. This may be the single biggest hurdle to reinventing education that AI makes possible, indeed demands, as it remakes our world. Yet it may be the single biggest opportunity to get education right.

This is not entirely unchartered territory. The *flipped or inverted classroom* model became very popular starting about 2010. Most flipped classrooms have students consuming learning outside of class, such as watching prerecorded lectures, reading, doing self-paced tutorials, and other learning activities. Subsequently, in the classroom, the teacher, abandoning the role of "sage on the stage," transforms class time into an intellectual forum for higher-order cognitive work, complex collaborative problem-solving, targeted Socratic debates, and real-time coaching. In the flipped classroom, teachers deploy their expertise not to transmit information, but to facilitate the difficult, often messy journey toward mastery and critical discernment, while working on the soft skills development prized in the workplace and in life.

The flipped model has probably found its most widespread use in science, technology, engineering, and math fields, pioneered by notable educators like Harvard's Eric Mazur, a professor of physics who championed the flipped classroom (though he prefers the term *peer instruction*). Mazur uses the model to provide time-on-task flexibility for students to grapple with sequential, difficult material at their own pace, while using class time for knowledge checks and then peer-to-peer work.[5] Like any model, the flipped classroom has its pros and cons. It raises questions about the digital divide, presupposing that all learners possess the requisite internet bandwidth, device access, and conducive home environment necessary for consuming video lectures, an assumption that can reinforce existing socioeconomic disparities. Furthermore, it is in some ways harder work for a teacher, who must not only produce high-quality instructional work for students' time outside of class but also manage a more dynamic classroom. While modestly successful, the flipped classroom might be another one of those artifacts from the future that foretells much more profound changes to come. With the rise in AI capabilities, a genius TA can provide far more robust learning than most of today's students can do on their own: providing faculty with detailed analysis on each student's progress and even lesson planning. Most important, class time can fully become *human time*, when laptops remain closed and AI is nowhere present (or at least as little present as a teacher wants). The classroom becomes a space reserved for being fully present with one another without the constant pull and distraction of technology. That will create a different kind of role for the faculty, who will inhabit a role that is more human, more relational, and far more central to the learning experience.

That does not mean faculty avoid spending time on the disciplinary knowledge that students have been mastering outside of class. The genius TA might flag a student struggling to master or apply a concept, recommending some form of intervention to the instructor. In a university computer science course, for instance, the genius TA would assume administrative and diagnostic tasks—grading routine problem sets, tracking time on task, and, crucially, pinpointing the *exact moment* a student's logical model breaks down, perhaps revealing a persistent error in understanding object-oriented inheritance. The professor then leverages this granular data not to cover the concept again, but to provide a precise high-impact insight for that *student*: a sixty-second personalized analogy that unlocks learning and is delivered at the student's desk. This shift from mass instruction to real-time targeted care ensures that the highest-leverage learning activities involve human interaction, those aha moments when an expert teacher can see things click for a student.

The powerful diagnostic capability of AI can provide a basis for establishing relevancy (what most students crave) and developing judgment, activities best managed when teacher and students are gathered in class. Consider an AI course on developing agentic AI and building safeguards. Using an opportunity to bring in philosophy, students might use a large language model (LLM) to synthesize and understand the core arguments of a complex text by someone like Immanuel Kant, thereby satisfying the low-order task of comprehension—but then in class be confronted with a modern adaptation of Kant's "inquiring murderer" thought experiment. The teacher can now use the actual class period to facilitate a discussion on the question, which can complexify and illuminate the topic at hand (building safe AI is *hard*), raise the issue of design and unintended consequences, foster critical thinking, develop discussion and debate skills, and use the whole as a forum for better knowing the students. A skilled teacher can build the crucial bridge between knowledge and wisdom in ways that are fundamentally human and deeply owned by the learner and the teacher.

Across campus, in the School of Architecture, a genius TA in a design class can work with architecture students to perform tedious, high-computation tasks—running stress tests, checking building code compliance, or managing complex material sourcing logistics. This effectively removes technical barriers to creative iteration while developing, accessing, and reporting on student skills in each. It frees the professor to focus less on technical competence and more on the higher-level questions that shape actual mastery: delivering the subjective, tacit, and experiential critique on *form, meaning, and cultural impact*. The teacher's intervention shifts entirely to the cultivation of discernment and ethical imagination—guiding the student not on *what* a machine can calculate, but on *what the design should mean* for the community it serves.[6] In partnership, the machine patiently handles mechanics, foundational skills development, and training, liberating the faculty member to focus more on cultivating the unique, complex sensibilities that only human consciousness can generate.

When we talk about a genius TA that can do some much of the foundational knowledge transfer, act as a nonjudgmental twenty-four/seven tutor, assess student progress, and make recommendations for classroom time activities, it can sound to a worried faculty member (and there are a *lot* of worried faculty members) that they are in danger of becoming obsolete. That is not the case, as the previous examples illustrate, outlining as they do a set of more important higher-order responsibilities for faculty. These include intervening in high-impact moments when a student is stuck and the AI tutor has not unlocked learning, and curating class time to lift students from merely knowing the material to applying that knowledge in varied scenarios that require critical thinking, judgment, and discernment. Faculty also can use class time for enrichment and developing the human skills that will be more important in the Age of AI, such as effective communication, constructive dialogue, navigating cultural differences, empathy, creativity, working in teams, and professional disposition. As important, faculty can use the time the genius TA frees up for building the genuine and deep relationships with students that we described at the outset, the kind that make them feel like they matter, that they have someone in their corner, the kind that fuels transformation.

The best teachers already do much of what is just described. But many, maybe most, still focus on knowledge transfer. The ones who *want* that kind of relationship with students and yearn to do the higher-order work of judgment, wisdom, and soft skill development often are overwhelmed with students who are ill-prepared for the basic work of mastering the discipline, burdened with heavy student loads, or pulled away for other myriad duties—or some combination of the three. The evolution of the faculty member's role will require three key developments:

- **Learning to work with an AI teaching assistant or partner:** No matter who provides the AI agent—the institution, a third-party provider, or one designed by the faculty member (not a small undertaking)—instructors will want trustworthy agents that they can shape in terms of how they work with students and that they can direct as needed. Along the way, they will learn how to work well with these tireless and intelligent assistants.

- **Shifting the focus of instruction:** Moving from knowledge transfer and assessment to the kinds of higher-order skills previously described will require faculty to evolve their pedagogy, what they do in class, and ensure that their own expertise teaching skills such as critical thinking, constructive dialogue, and performance-based assessment are up to the task. Centers for teaching and learning, which generally support faculty development, will be busy places in the years ahead.

- **Learning to love students again:** So much of higher education works against a teacher's ability to develop deep, transformative relationships with students. In many institutions, the recognition and reward structures favor research and scholarship (those rewards usually mean *less* interaction with students, not more). In other kinds of institutions, faculty loads and class sizes mean little time to get to know students. Most faculty I know care deeply about students, yet only a small percentage of them have the time to know maybe more than a few in any meaningful way. A new paradigm that asks teachers to embrace a relational mode of engagement will require for many faculty the exercise of little used muscles, a challenge much more about mindset and habits of the heart than about actual pedagogy. When I described this shift, a longtime colleague grumbled, "I signed up to teach them political science, not be their counselor or friend." He has some work to do.

It would be tempting to ask, "How do we, then, have to rethink the training and preparation of faculty?" We might first point out that the great majority of faculty members received little to no training or preparation on how to teach. We will need to be more intentional in our preparation of faculty entering our classrooms and retraining the many thousands of faculty who already teach in our universities. This will be postsecondary education's own version of the *reskilling* challenge that will affect every industry and job in our society.

It is important to note that the focus here is on teaching and learning. More sections of university courses are taught by adjuncts or contingent faculty than by full-time faculty, a trend that started decades ago. They are asked to do nothing but teach, but full-time faculty do many other things. The life of a faculty member, particularly at a research-intensive institution, has never been about teaching alone. The traditional triad has involved teaching, research, and service. Having focused on the transformation of teaching that AI will make possible, we should take a moment to touch on the other roles that full-time faculty are often asked to play.

The first and most defining nonteaching role is the *scholarship and research* function, a commitment to the creation or at least extension of original knowledge. In the pre-AI era, this role was defined by a massive expenditure of mechanical labor: the laborious sifting through archives; the tedious, often repetitive tasks of manual data processing; the exhaustive bibliographic searches required to map the contours of a given field. The machine, now unchained, performs this scaffolding labor with superhuman speed. An LLM can process the entire body of specialized literature—a feat requiring decades of human labor—in seconds, identifying hidden correlations, synthesizing complex summaries, even writing the first mechanical draft of a methodology section. Yet the human scholar retains intellectual sovereignty. The essential scholarly task is transformed from the *gathering* of information to the discernment of meaning. Humans must still ask the relevant and generative questions, see what is missing, frame problems in ways AI cannot, and think across disciplines to recognize patterns that lie outside any single domain. They must exercise judgment, decide which lines of inquiry matter, interpret findings in context, and anchor research within ethical, social, and historical horizons that machines cannot adequately grasp. If the teacher of the future will partner with a genius TA, the scholar of the future will partner with some kind of genius research assistant. Even a poorly supported scholar in a non-research-oriented institution has become uber-powered through access to AI.

In a genomics lab, a researcher might use AI to process petabytes of sequencing data to flag genetic variants. But it is the researcher who must formulate the hypothesis, asking, "Which variants are significant? Which correlations warrant the profound cost of clinical intervention? Are there biases at play here?" In history, an algorithm can sift through millions of colonial documents to reveal patterns of trade or migration, yet the scholar's value resides in providing the interpretation and moral context—in asking *why* these patterns persist and *what* they mean for contemporary justice and remembrance. The professor must become a curator of intellectual courage, unafraid to challenge the subtle biases embedded in the machine's training data, which in the end reflects mostly what is known and what we, as humans, have created (for good and, sometimes, bad).

There may very well come a time when AI-driven or -guided research extends human knowledge well beyond that which we can understand (if we create AI "super intelligence" combined with quantum computing, for example). But for now, AI is a powerful extension of our cognitive powers that enables us to ask new and better questions—questions that AI, trained on what we know rather than what we might imagine, cannot yet address. AI has supercharged our research and scholarship, but humans must direct and shape the work, and our creativity is required to open new fields of inquiry, to answer more interesting questions, and to solve seemingly intractable problems. They must do so while determining which questions matter for the communities they serve, which avenues of inquiry are worth the investment of time and resources, and surfacing the ethical and moral dimensions of the work itself. It is not clear that research will be much linked to teaching, though in many or even most institutions, that is already largely the case. Research faculty are much more likely to work with graduate students and have minimal teaching loads.

After teaching and research/scholarship, faculty have a third less glamorous, yet important role in institutional service and stewardship, the essential hidden labor required to maintain the university as a self-regulating human institution dedicated to intergenerational continuity. These are the endless meetings: the curriculum committees debating the inclusion of a new quantum computing module, the tenure boards deliberating the fate of a promising assistant professor, the policy committees drafting the university's code of conduct for generative AI. While AI can certainly schedule the meetings, draft the minutes, and even predict the outcome of a departmental budget vote, it cannot perform the central nonalgorithmic function: ethical judgment and political discernment. The university is not a business; it is a moral enterprise built on trust, intellectual rigor, and shared purpose. As universities come to declare their values, their commitment to character formation, and rediscover their role in creating good human beings and good citizens, they will need the faculty to shape the moral landscape of the institution.

I have a confession to make. There have been times during the twenty-eight years I served as president of two very different institutions when I questioned the value of tenure and shared governance, associating the former with simple job protection for even the most incompetent and poor performing instructors and the latter with glacial processes ill-suited for modern organizations in a world moving at lightning speed. I would argue that both are often true and that very few presidents would disagree if they could speak freely. Yet in this current climate of rising authoritarianism, Orwellian post-truth propaganda, and assaults on academic freedom by ideologues (mostly on the far right, but also for a long time by those on the left), we need set the intellectual compass of the university using evidence, rational debate, and scholarly research as lodestones in our increasingly post-truth, anti-expertise, and ideologically shaped society.

While those needs are mostly enacted within the university and the disciplines, there is in a civil society another increasingly important role for faculty, that of the public intellectual and moral steward. This is the unwritten contract of academic life: to act as the primary translator between highly specialized knowledge and the pressing, often fragmented needs of society. The professor steps out of the classroom and the laboratory and into the public sphere—writing the essay that contextualizes a new pandemic, testifying before a legislative body on the ethical risks of facial recognition technology, or modeling reasoned debate in an age defined by polarization. This role has never been more vital, because if the Age of AI is fundamentally an Age of Information Overload and Opacity, the world desperately needs trusted arbiters of meaning.

We live in a time when some combination of ideology, politics, and commercial interests have eroded society's trust in science and expertise. This has been vividly illustrated in the vaccine debates, the gutting of America's public health and research apparatus, and near-absurdist appointment of Robert F. Kennedy Jr. as our secretary of the US Department of Health and Human Services. If we are to rebuild an evidence- and science-based approach to solving the complex challenges comforting us as a society, we need our universities and our scholars and researchers to be leaders in the effort, the foundation of that effort. They must position the university as a moral and intellectual commons, a space where intellectual courage is modeled, where complex problems are debated *horizontally* across disciplines, and where the goal is not consensus but clarity of thought. As machines increasingly enable us to address complex issues and drive breakthroughs in climate change, medicine, energy, and warfare, the faculty's function will be teaching how to understand that work and to support, cultivate, even lead the important public debates about what do with the what the technology and research make possible. We need our faculty to ground our progress in our humanity, ensuring that as machines excel at calculation, human beings excel at the more difficult but necessary work of meaning making.

In a sense, our world needs faculty members more than ever to play a role that has not been generally asked of them in a long time. It requires faculty not to do less, as AI covers more of their tasks, but to do more—and not just how they rethink class time, work with students, or forge relationships with students. In a world in which we cannot trust AI companies to work for the good of society or humanity, and our politics are deeply broken, higher education becomes one of the last social institutions capable of cultivating the habits of mind a democracy requires. This means *faculty* must serve as highly specialized experts in human judgment, ethical reasoning, and the cultivation of discernment. They must help students learn how to weigh competing truths, evaluate evidence in context, and recognize the moral stakes embedded in every technological and scientific advance. In this future, the university's value is secured not by its ability to compete with AI on speed or scale, but by its unwavering commitment to what the machine can never replicate: the human work of generating, governing, and contextualizing knowledge for the ultimate purpose of advancing a just and flourishing society. The Age of AI calls for reclaiming an elevated societal role for universities; it also calls for a commensurate elevation and expanded role for the faculty.

Reimagining the Postsecondary Ecosystem

For all the ways we can get AI wrong, particularly in education, it offers a thrilling invitation to harness its potential to enable a powerful model of education that is not only commensurate with the new Age of AI era we are entering but also necessary, even imperative, if we are to build a new world order that keeps humans and humanity at its center. The educational model outlined in the preceding chapters imagines a profound shift from the Information Age economy to a new care economy built on human flourishing, the health of communities, and the kind of new Golden Age that comes after every technological revolution. We do not today educate for that world.

Doing so requires shifting from the epistemological to the ontological, from content delivery to the intentional cultivation of human skills, from teacher as expert to teacher as curator of human connection. A care economy focused on human flourishing would look different than what we mostly see around us. It would offer amazing public schools filled with talented teachers, coaches, social workers, and staff, places where kids feel loved and nurtured and are enthralled with learning. It would mean robust, compassionate systems of care for our most vulnerable: the elderly, the poor, and those struggling with mental health and addiction (and a healthy society should have a lot less of both). It would be a place in which the arts are everywhere, and each person's creativity finds outlet and expression, as was once true of earlier societies and nowhere evidenced in our developed modern economies. Care would extend to our environment, since any Golden Age must include a planet not hostile to human life, as ours is increasingly so. This imagined future can sound naively utopian, but what is the point of our work as educators if not to realize such a vision?

Higher education can rise to this moment. It must to remain relevant in the Age of AI. We remain in the turbulent transition from the "installation period" of AI, dominated by financial speculation and chaotic disruption, and not yet in the "deployment period," characterized by shared prosperity, institutional reform, and a renewed social contract. The ugly transition in which we find ourselves has all the hallmarks of installation periods: huge wealth inequity, monopolies, populism, splintered politics, the rise of authoritarianism, an increase in wars. Yet these periods have always functioned like earthquakes after which society rebuilds in wondrous new ways. Higher education is in the midst of an earthquake, and the commonsense question is, "How do we get through this?" Survival is our individual and institutional imperative. Any future model of education must account for the economic, policy, and regulatory contexts in which the reinvention of education must take place. Even as we find ourselves in the earthquake, it is not too early to start imagining what we must build when the shaking stops. Our response must be as audacious as the AI revolution itself if we are to transform a brilliant invention into the foundation for a flourishing, equitable society.

Financial Bridge to Care Economy

Before turning to education and the changes needed to support its transition for the Age of AI, it is helpful to briefly address how a care economy can possibly come to life. The economic argument, introduced in chapter three, is predicated on the belief that AI will displace a massive number of knowledge, information, and technical jobs, requiring humans to pivot to high-touch relational professions (teaching, nursing, social work, elderly care) where empathy and judgment are paramount. Yet an economic truth remains stubbornly in the way: society historically undervalues and underpays for the very jobs we desperately need to create meaning, community and social stability (and, not coincidentally, many of those jobs are done by women). We reward efficiency and abstraction; we penalize presence and care. Record profits are privatized, but the societal costs—job displacement, social instability, and the need for reskilling—are socialized.

When I have shared this vision for a care economy, the very first question is always, "How will we pay for it?" Financing human infrastructure is the primary policy challenge of the Age of AI. We cannot shift from a high-status, high-salary knowledge economy to a low-status, low-salary care economy. That would be a failure. To recover a stable middle class, with more people doing more meaningful work rooted in relational value rather than information processing value, care economy jobs must be well supported. And that requires a new system for wealth redistribution, as argued earlier. The massive wealth being created by the core AI industry—the trillions captured by chip manufacturers, large language model developers, and cloud providers—cannot remain siloed. The engine of the new economy is generating vast value while simultaneously dismantling the job market of old. Nvidia, OpenAI, Google, Amazon, and others are capturing economic value at a scale unseen since the industrial barons of the nineteenth century, yet their products are designed to radically reduce the number of human hours required for creation and labor. This economic anomaly demands a policy response aimed at wealth redistribution and, ultimately, *universal human investment*.

In the past, economic prosperity was taxed through income and labor. But what happens when the labor is cognitive and automated, performed by algorithms that do not earn a salary, but whose output generates immense profit? Bill Gates was not the first to propose a tax on the robots. But his articulation of the idea is compellingly pragmatic: if a robot takes fifty thousand dollars of work away from a human worker, it should be taxed similarly. We must redefine the tax base. If wealth generation is the new era is rooted in massive automated cognitive power and in the enormous energy- and water-guzzling data centers that house it, then that is where the tax base must shift. This is not punitive, but rather the structural mechanism by which the profits of efficiency fund the needs of humanity. We are not taxing innovation, but the automation of labor, ensuring the fruits of that automation are reinvested in human capital. The specific policy levers are clear: a tax on the transactional value of automated cognitive labor; a carbon or water consumption tax on the massive data centers required to run advanced large learning models; or a temporary wealth tax on the largest valuations derived directly from AI automation. The creation of a stable, sustainable financial foundation for the care economy does not undercut or slow our advance into the Age of AI, it ensures we save AI from itself, just as President Franklin D. Roosevelt's social policies in the 1930s saved capitalism from its own excesses and possible demise.

This funding must be explicitly and transparently directed toward two goals:

1. **Elevating the care economy workforce:** Directing resources to provide the salaries and training of teachers, home health aides, social workers, counselors, artists, and others to transform those jobs into secure, respected, and well-compensated professions must become the new engine of the middle class, replacing the jobs lost in middle management and information processing. It would serve as a public acknowledgment that *caring* is essential, complex labor that benefits the entire community.

2. **Making high-precision learning universal:** Ensuring that the AI-powered education system, which is fundamentally cheaper at the margin due to the automation of content and assessment (as noted in chapter five), is either free or radically affordable for all, regardless of age, background, or geographic location. We must become a learning society.

It is important to point out that this is not the universal basic income (UBI) model so often tossed out as a solution. UBI is not as much about flourishing as providing just enough funding to keep a population from revolting. It does not usually end well for society when humans lack *meaningful* work. As argued earlier, care economy jobs are far more meaningful than most in the finance, technology, and information economy sectors *if* we pay enough, support the people doing them, and afford them respect.

This financial shift recalls one of the great societal successes of the last century: the post-World War Two investment in education. The G.I. Bill was a benefit for veterans, but more importantly an unprecedented act of public financing that transformed higher education from a privilege into the engine of the American middle class. It funded the *who* (the students) and, by extension, the rapid expansion of the *what* (the universities), creating a generation of highly skilled, educated citizens who fueled decades of economic boom. It enabled a new version of America in 1955 that scarcely resembled the America of 1935.

The Age of AI requires a similar commitment not just to the *student* but to the entire *ecosystem* of care and learning. We must move from an education model focused on scarcity (limited seats, high price, limited faculty time) to a model focused on abundance (infinite content, personalized support, affordable access). AI-powered learning, with its genius TA and precision-learning architecture, makes the vision of universal access technically feasible; the new tax architecture makes it financially responsible. With scalable, low-cost access comes the opportunity to reinvent the roles of the classroom, the faculty member, and education overall in the ways earlier argued.

We must use the abundance generated by technology to fund the human flourishing that technology cannot replicate. Failure to do so condemns us to an ever-widening canyon of wealth inequality, fueling the social discontent and political instability that Perez warns is endemic to the "installation period."

Rethinking Credentials

The traditional college degree, a single, static credential awarded after a fixed period and based on a fixed set of inputs, is a relic of the Industrial Age. Its design assumes two things that are no longer true: that knowledge is scarce and mostly static, and that one's career is a relatively fixed path lasting forty years, built on an initial foundation of a two- or four-year degree. Even before AI burst on the scene, those notions were breaking down, evidenced by the rise in micro-credentials and short-form non-degree options for learning (from badges to certifications from companies like Google, IBM, Salesforce, and Oracle), the increasingly short half-life of skills requiring near-constant reskilling, the average number of career changes or projected for new entrants into the workforce, and the rise of new forms of work for which traditional degrees seem unaligned (from gig workers to influencers and digital nomads). The traditional degree, long designed to be the final word on one's education, is becoming less valuable the moment it is printed, almost the way a new car drops in value when driven off the lot. While the degree remains important in other ways—its status signaling, its existential reassurance, its coming-of-age stamp of approval—it is, from a workforce perspective, a snapshot in time, a photograph taken as a learner walks off the graduation stage.

The traditional degree will need to give way to a much more dynamic, constantly updated picture of any individual's skills and attributes, a digital learner profile that captures credentials earned: degrees, micro-credentials, certifications, verified skills, and myriad other attributes a person wants to represent to the world. More than a digital transcript, it is a continuously updated evidence-based map of all abilities. Unlike the degree, which focuses only on inputs (courses taken, credits earned) and a single institutional output, a transcript, the learner profile requires evidence and validation of mastered capabilities. It tracks three crucial, distinct categories of human capital: mastered skills, specific technical abilities or foundational skills, such as a nurse's ability to understand and work with human bodies; core competencies, transferable intellectual skills like critical thinking or quantitative reasoning; and, most critically for the Age of AI, ontological capacities—deep, human attributes like judgment, empathy, and collaboration. There are attempts underway to better capture a student's array of skills and experience beyond courses completed and grades earned. Systems like the Comprehensive Learner Record, adopted by the University of Maryland, are replacing transcripts with records that include skills mapped to external frameworks like the National Association of Colleges and Employers' competencies.

The velocity of change in society and the in modern economy demands that we move from a fixed measure of educational input to a dynamic, verifiable high-resolution signal of current human capability, one that aligns with an educational ecosystem that accommodates far more forms of credentials and learning, from a wider variety of courses, and with greater granularity—all happening across a career and a lifetime. Learning is becoming a lifestyle, not a time-boxed activity in and of itself. The learner profile provides longitudinal currency; it is an asset that appreciates and adapts with the learner's ongoing professional and human development, reflecting that the pursuit of expertise is now continuous, not finite.

This fundamental redefinition of learning shifts the entire credentialing focus from completion (Did the person finish the fixed course of study?) to competency (What can the person do right now? How have you verified that capability?) and growth

(What do you need to master next and why?). The learner profile as imagined here also reflects an end to higher education's eroding monopoly on what counts as learning. The knowledge transfer function of learning is decreasing in value, and that will be accelerated as AI provides amazing learning fine-tuned to the individual and offered at an extremely low cost. As a result, the role of universities will shift from providing knowledge and skills to assessment and validation, conferral of trusted certifications, and development of human skills, as discussed previously. In such a world, the transcript and diploma, owned and conferred by the university, gives way to the learner profile, owned and managed by the individual and validated by the university.

The single-most important shift in credentialing is moving from time-based units (seat hours, credit hours) to evidence-based mastery of skills or competencies. The time required to learn a skill is highly individualized, and a system that punishes students who take longer or forces others to wait for the average pace is inherently inefficient and inequitable. As I explored in *Students First: Equity, Access, and Opportunity in Higher Education*, the credit hour is a terrible basis for measuring learning, better at telling us how long one sat than what one learned. It is why Southern New Hampshire University (SNHU) launched College for America, a competency-based model that was the first to untether learning from time.[1] Time-based models are inequitable: low-income learners have less time and less control over their time than those with privilege. The ability to spend two or four years concentrating full-time on a degree is a luxury afforded to a diminishing number of learners. A new system of education will focus on measurable skills, a combination of what you know and what *you can do* with what you know, completed at a pace that fits the student's need and context.

In a world where constant upskilling and reskilling is the norm, degrees are like battleships in a sea of speed boats:

powerful, expensive, but lumbering. The new education ecosystem will include these elements:

- **Modular stackable credentials:** The future curriculum will be broken down into granular skill-focused units: micro-credentials that can be stacked, aggregated, and rapidly validated by institutions and industry. A graduate may hold a mastery certificate in ethical AI deployment alongside a foundational knowledge badge in geriatric nursing or a certificate in intercultural conflict mediation. This agility enables workers to quickly reskill and reenter the workforce, a necessity when a job can be redefined or eliminated in eighteen months. The university's program portfolio becomes a set of interlocking competencies, not a fixed course catalog, and people will become accustomed to moving in and out of a broad and diverse educational ecosystem as they need or want.

- **The dynamic portfolio:** The degree must be replaced by the living, continuously updated learner profile that dynamically and reliably tracks ongoing learning, whether it is four hours of training, four days of practicing a skill, four weeks of concentrated study, or four years of a degree program. Powered by the AI's real-time data ingestion (as noted in chapter five), this profile tracks mastery against knowledge graphs, recording not just *that* a student took a class or a program, but *when* and *how* they demonstrated competence in a specific skill and at what level and in what context, insight not provided in traditional transcripts. The learner profile documents the full journey: the projects completed, the peer endorsements received, the simulated ethical crises navigated, and the growth trajectory of their human skills. This is the new currency of learning, instantly transferable and perpetually current.

While no single taxonomy has come to inform skills assessment, higher education is getting better at defining and measuring lost skills as the competency-based education movement continues to grow. As education increasingly shifts to more focus on the ontological dimensions of learning, we will need to get better at defining and assessing those skills. It is one thing to give students a grade in a history course; it is an entirely different matter to certify that they possess demonstrable ethical judgment or intercultural communication competence. This challenge requires a break from traditional subjective measures, embracing tools that bring quantitative rigor to qualitative skills.

We are not starting from scratch. Fields like medicine, nursing, and teacher education have long maintained rigorous, high-stakes assessment practices designed to certify ontological capacities and complex human performance skills. Clinical rotations, practicums, and residency evaluations already function as real-world performance-based assessments. A student nurse's empathy, professional judgment, and collaboration are evaluated not in a lecture hall, but through the direct observation of their interactions with patients, family members, and medical teams by experienced preceptors. The use of the Objective Structured Clinical Examination, where students rotate through simulated patient encounters assessed against standardized rubrics, demonstrates that certifying human skills through structured, real-time observation is an established, high-stakes practice. This existing framework provides the proof-of-concept for extending rigorous performance-based assessment far beyond the clinical fields.

We will need to use technology and other assessment approaches to scale and certify human skills. These might include the following:

- **Performance-based simulations:** We can use AI-powered agents and environments to place students in complex scenarios that assess actions, not memorized facts.

- **In medicine:** A nurse's assessment is based on the ability to handle a simulated emotionally charged conversation with a patient who refuses treatment, where metrics capture not just what they *say*, but their *tone, listening behavior* (via voice analysis), and *adherence to compassionate care protocols.*
- **In engineering:** A student is graded based on the capacity to lead a diverse virtual team to resolve a supply chain crisis. The AI tracks team communication patterns, conflict mediation success, and ethical trade-offs made under pressure.
- **In social work:** A student simulates a conversation with a role-playing AI agent to practice deescalating a tense family conflict, with assessment rubrics focused on active listening, nonjudgmental language, and successful rapport building.

These tools exist today. Medical schools and nursing programs use high-fidelity patient simulators (mannequins that react physiologically) and virtual reality simulations for surgical training. Beyond clinical, platforms like Mursion use human-in-the-loop avatars for teachers and managers to practice high-stakes conversations and conflict resolution.

- **Three hundred-sixty-degree and peer feedback:** This process would incorporate peer, mentor, and employer feedback into the official record. Students' collaboration skills should be assessed by the team with which they worked on a complex capstone project using standardized, validated rubrics for interpersonal dynamics. This grounds the assessment of human skills in a real-world relational context, moving the assessment of these skills from the abstract to the accountable. I chair the Academic Council of Paris-based Forward College, which uses a three hundred-sixty-degree evaluation for students every semester to help them develop their human skills of communication, collaboration, and leadership.

- **Longitudinal evidence (the trajectory):** The system must record the *growth velocity* of a human skill. A student does not simply *acquire* empathy but rather *develop* it over time through intentional practice. The credential must document that trajectory, including moments of initial failure followed by evidence of improvement. This moves the assessment from a single point in time grade to a demonstration of learning and adaptation over time.

The university's role as the certifying body is territory to be reclaimed. Employers no longer feel they can rely on the labor market signal offered by a degree. They have hired too many college graduates who cannot write well, do not possess basic quantitative skills, or know how to show up in the workplace. Decades of grade inflation, outdated academic programs, and poor outcomes have eroded trust in higher education's ability to validate learning and skills. The sector will have to regain that trust with more rigorous assessment practices, a focus on outcomes and not inputs, and transparency for the claims we make about our graduates. The value of new credentials, however they take shape, will not be in the technology that drives them, but in the trust with which the university infuses them. When hiring managers see a credential certified by a university that uses rigorous evidence-based human skills assessment, they can be confident they are hiring a capable human being, not just a fact-recollection machine.

Rethinking Accreditation

The accreditation of institutions has been under attack for some time now. While few dispute its core need—assuring the public, students, and employers that an institution meets sufficient standards of quality—it has become a battleground in the United

States, with harsh critics from the left ("You allow low-quality and disreputable providers, especially for-profit providers that prey on students who are then burdened with excessive debt and a low value credential, if they finish at all.") and the right ("You protect the status quo and an existing monopoly, a kind of regulatory capture by the industry, while smothering innovation and overreach into areas not yours to regulate."). There is truth in each perspective, and good reason for institutions to chafe at accreditation given the cost, bureaucracy, and accountability each imposes. There is hunger for new approaches, but little agreement on what those should be.[2] At the time of this writing, the Trump administration is finding ways to disempower the mainstream regional accreditors, pressure them to enforce its own political agenda for things like diversity, equity, and inclusion programs and approve new accreditors.

The current debates are about renovating the house, what exists today. We need a new house for the new era of education. What might accreditation look like?

An emerging new ecosystem of learning, as just described, with the paradigmatic shifts elucidated in previous chapters, requires accreditation to move away from institutional inputs and time-based measures to a rigorous, continuous validation of evidence-based outcomes. It must pivot from being a periodic bureaucratic check to a dynamic, continuous certification of the credentialing engine itself. The central question it must ask is, "Can this provider's certification of learner performance be trusted?" This new approach would focus on three main shifts:

- **The shift from input-based review to outcome-validation:** The current system spends undue resources reviewing physical assets, faculty degrees, and financial ratios (inputs). The new model must focus on the integrity of evidence and assessment. For example, an accreditor would not primarily review the institution's curriculum or clock hours, but rather the learner profile's knowledge-graph architecture, including how it is validated, presented, and assessed. The review would certify that the university's systems reliably and fairly define, measure, and validate the three categories of human capital: mastered skills, core competencies, and ontological capacities. This shift would include a much greater focus on assessment. That might include auditing an institution's use of performance-based simulations or objective structured clinical examinations to ensure the rubrics are standardized, the AI-driven metrics unbiased, and that the assessments genuinely measure judgment and empathy, not just memorization. As we increasingly use AI in role-playing, simulations, and other forms of assessments, accreditation review will need to become expert in agentic design and evaluation (Does the system do what it is intended to do?). Finally, the accreditor would ensure the digital security, immutability, and chain of custody for the evidence (e.g., project data, peer feedback) that flows into the learner profile, maintaining trust in the high-resolution signal provided to the labor market. In the Age of AI, when we no longer trust videos, voices, texts, or other forms of digital information, it is hard to overstate how important this aspect of examination, assessment, and certification is about to become.

- **The shift from time-based audits to continuous monitoring:** The traditional cyclical, time-intensive audit is obsolete in an era of continuous reskilling and fast-changing job requirements. The typical accreditation cycle is an extensive review every ten years, with a more modest check-in review at the five-year mark—a five-year cycle in a world changing at light speed. That obviously needs to change, and we should be thinking about these issues:

- **Continuous compliance monitoring:** Using technology, the accreditor should have a system of real-time data feeds to monitor key performance indicators related to student success. This could include a skills-to-employment fit, which tracks the correlation between certified skills in the learner profile and employment placement/salary data, focusing on the longitudinal currency of the credential; and periodic growth trajectory audits that gauge the institution's capacity to document longitudinal evidence of human skill development, ensuring the system measures growth velocity and adaptation over time rather than a single point in time score.

- **Agility for modular credentials:** The process must be agile enough to certify modular, stackable credentials (micro-credentials and certificates) rapidly, enabling institutions and the other new providers entering the ecosystem to quickly launch high-quality reskilling programs without waiting years for approval. This supports universal high-precision learning that is affordable, accessible on demand, *and* also highly responsive to changing labor market needs.

- **The shift from protecting monopolies to certifying trust in the ecosystem:** Traditional higher education's monopolistic hold on what counts as postsecondary education has been slipping for some time. New providers have emerged in myriad forms, including boot camps of various kinds; massive open online course providers such as Coursera and edX, which have morphed into outsourced program managers and delivery platforms; corporate offerings such as Grow with Google and Trailhead from Salesforce; and new technology-powered, and increasingly AI-powered, "universities." The large AI companies themselves are moving very quickly into learning. Students have a vast, growing, and often bewildering array of places they can turn to for learning, creating a messy, emerging ecosystem.

Today's accreditation standards give scant attention to the core characteristics of learning outlined in this book: the shift from knowledge transfer to human development; precision, or at least personalized, learning; holistic support for student well-being; rigor of assessments; the centrality of technology generally and AI in particular (with all the issues of data privacy, governance, and security that get amplified); and all of the ways and places in which students might learn and how the new learning ecosystem assesses and accommodates them. The enormity of these changes will require the kind of bold, imaginative rethinking not often seen in current accreditation debates.

Access and Equity

Over my twenty-eight years as a college and university president, I was most moved in the places where our students have the least. I remember sitting in a dilapidated building with dirt floors and rough wooden benches in the Kiziba Refugee Camp, near the shores of Lake Kivu in Rwanda, in 2014. It was densely packed, full of young refugees who had been born in the camp and never experienced life outside of its near-desperate desolation. We met with ten young refugees, all graduates of the camp's primary school, and as we talked about their education, their next steps, there was a kind of hopelessness in their eyes. The questions felt almost cruel given the lack of prospects or opportunities. When we described our idea for bringing SNHU's online degrees into the camp, they nearly jumped out of their seats, bombarding us with questions about how soon, what programs, and where to sign up. Through the efforts of a dedicated team, generous donor startup funds, and a partnership with Kepler University in Kigali, that meeting led to the creation of SNHU's Global Education Movement or GEM Program, which has come to educate displaced peoples in camps in other locations in Rwanda, Malawi, Kenya, South Africa, and Lebanon. It may be the work of which I am most proud in my twenty-one years at SNHU.

When I described our refugee work to former US Secretary of Education Arne Duncan, who has dedicated himself to combatting gang violence and creating opportunity in Chicago, he asked, "If you can do it in Rwanda, can't you do it in the South Side of Chicago?" That led to our Community Partners programs, which brings programs into underserved communities across the country. As I saw with learners in the camps, earning a degree was transformative for our graduates, creating opportunity, hope, and self-esteem. SNHU loses money with these programs. As the leader of a nonprofit whose mission comes first, I could always defend those losses and, truth be told, no one ever really questioned them. This work made our employees proud, and our board of trustees asked, "How much more do you need?" not "How much are we spending?" This has always been my driving passion: to bring the transformative power of education to those who are underserved or for whom there simply are no workable options.

AI can be the game changer in creating sustainable, radically affordable, and scalable education to millions of people around the planet. The Matter and Space platform we were building was designed to do just that, as we worked to get the cost of delivery to under thirty dollars per month. At that price point, our platform, delivered on low-cost devices and with approaches that minimized bandwidth and access challenges, could allow new delivery models and truly democratize education, addressing a global crisis of access and relevance. If AI is to create a Golden Age, it cannot be locked behind the gates of the wealthy world. This is not charity; this is also a matter of global security. Uneducated, underserved populations are far more vulnerable to misinformation, exploitation, and the destabilizing forces of unchecked technology.

The ultimate moral test of the AI revolution is whether it exacerbates the global education gap or closes it. Low-cost, high-precision models, made possible by platforms like Matter and Space and architectures that separates content delivery (cheap, automated) from human mentorship (high-value, targeted, local and thus trusted), provide the only plausible path to meet the global demand for quality education. Imagine the expertise of a world-class tutor and the structure of a dynamic curriculum delivered to a refugee in a camp, a rural student in sub-Saharan Africa, or a working parent in a developing nation not through expensive, sprawling physical plants, but through an AI-powered ecosystem anchored by human mentorship. This is not simply a business model; it is a geopolitical imperative. A more educated, human-skilled, and engaged global citizenry is the ultimate guardrail against the destructive forces of polarization, instability, and the unchecked power of global AI monopolies.

The new social contract for higher education, therefore, is not a document written for the university alone. It is a pact with society: in exchange for public trust and necessary financial investment, the university will transition from an inefficient, inequitable engine of knowledge transfer into a global system of human development, committed to equipping every person with the wisdom, skills, and ethical compass needed to navigate and shape the Age of AI. We must rebuild the financial scaffolding to fund human work. We must redesign the credentials to certify human capacity. And we must orient governance toward ethical, equitable access. Only then will the technology serve the whole of humanity, not just the privileged few who invented it. This is the promise of the "deployment period," and the responsibility of every leader in the face of the technological storm.

Why This Feels Almost Impossible, and Why It Is Not

It is easy to arrive at this point in the argument, throw up one's hands, and decide that this is too much system change with too much complexity and impossibly high hurdles to clear for the vision to ever become a reality. The list of difficulties is long:

- There are still those who argue that AI will not displace jobs as much as redefine them and create new ones, as with other technological revolutions.

- Wealth redistribution would see fierce resistance from companies realizing record profits and from the increasing number of billionaires who exert influence through politics, media, and manipulation.

- AI is still too mistake-prone and unpredictable to be the foundation for knowledge transfer, and we still do not really know if it will remain inexpensive.

- The learner profile, with its comprehensive data collection and deep insights, is too close to surveillance technology, posing all kinds of risk in privacy, data security, and abuse by bad actors or the state.

- Higher education institutions, including accreditors, are too set in their ways and too slow to change for us to realize this vision any time soon.

If we were in a period of incremental change, these might be more persuasive. But we are fifty years into a technological revolution that has already transformed the world, with AI poised to provide the most disruptive chapter yet. Keep in mind that the Magnificent Seven—the technology companies that dominate the economy: Microsoft, Apple, Nvidia, Amazon, Alphabet, Meta, and Tesla—did not exist fifty years ago. Much of what we take for granted today would have seemed impossible then. What seems impossible at this moment may be more possible than we can know.

We have signals all around us that act as at least partial counterfactuals to that list of objections. A 2025 poll showed that 71 percent of Americans fear that AI will cause permanent job loss now or in the future.[3] The pace of improvement in AI, with fewer hallucinations or mistakes (nearing 98 percent accuracy in some models, better than most people I know), and its often uncanny capacities, are concrete reassurances that it is indeed getting good enough to offer high-quality instruction. The challenges of security and privacy are real, but they already exist and this will be a never-ending race in a world where so much of our data is already "out there." Policy will be important, but we will have to turn to technology solutions to address data security. Higher education is exceedingly slow to change, but crisis often induces dramatic and rapid evolution. And higher education, including accreditation, is in crisis. Finally, the market will drive change. Already, most online searches for postsecondary learning are for *nondegree* options, and millions of people are turning to new sources of learning.

Concerns over an AI economic bubble, rising fear or at least skepticism about AI, the likely AI disasters still before us, stepped up regulation, and other factors likely will make the transition period in which we find ourselves messy, uncertain, and painful. History tells us that these transitions are never easy. Eventually, Carlota Perez reminds us, things break and we must then ask ourselves as a society, "What does a good life look like?"[4] In higher education, our version of that question might be, "What does higher education look like for the new world order? What's our role in creating that good life?" It feels like a worthy goal for the future of higher education: where learning is everywhere and available in whatever way you need it, where it is affordable and abundant and more fully focused on creating healthy human beings in healthy communities, and where we are equipped to flourish in a world in which the alien intelligence of AI works and "lives" alongside us in beneficial ways. It's a North Star I'll embrace, even if the path there is uneven, fraught with danger, and often uncertain or confusing.

I am struck by the number of meetings I attend with current and former university presidents in which there is a tacit and sometimes explicitly stated belief that the game is over. At a recent dinner, the retired president of one of the most selective and elite universities in the country was asked by another guest, "What will universities look like when my fifth-grader graduates from high school." She replied, "I have no idea, but they can't look and work as they do today." We all know the old dispensation is hollowed out and something new is emerging, but we are too much in the middle of it, too much in the birthing of the new, to say with certainty what it shall be. I am reminded of one of my favorite poems, T. S. Eliot's 1927 "Journey of the Magi." The speaker is one of the three Wise Men, now near the end of his life, looking back and telling how they followed a star through strange lands to attend a humble birth in a manger, which they never really understood (and were even a little overwhelmed by). But forever after, they knew the old familiar world was done, bitterly so for them, even as they could not yet understand the world to come or the signs all around them:

A cold coming we had of it,
Just the worst time of the year
For a journey, and such a long journey:
The ways deep and the weather sharp,

The very dead of winter.
And the camels galled, sore-footed, refractory,
Lying down in the melting snow.
There were times we regretted
The summer palaces on slopes, the terraces,
And the silken girls bringing sherbet.
Then the camel men cursing and grumbling
And running away, and wanting their liquor and women,
And the night-fires going out, and the lack of shelters,
And the cities hostile and the towns unfriendly
And the villages dirty and charging high prices:
A hard time we had of it.
At the end we preferred to travel all night,
Sleeping in snatches,
With the voices singing in our ears, saying
That this was all folly.
Then at dawn we came down to a temperate valley,
Wet, below the snow line, smelling of vegetation;
With a running stream and a water-mill beating the darkness,
And three trees on the low sky,
And an old white horse galloped away in the meadow.
Then we came to a tavern with vine-leaves over the lintel,
Six hands at an open door dicing for pieces of silver,
And feet kicking the empty wine-skins,
But there was no information, and so we continued
And arrived at evening, not a moment too soon
Finding the place; it was (you may say) satisfactory.
All this was a long time ago, I remember,
And I would do it again, but set down
This set down
This: were we led all that way for
Birth or Death? There was a Birth, certainly,
We had evidence and no doubt. I had seen birth and death,
But had thought they were different; this Birth was
Hard and bitter agony for us, like Death, our death.

We returned to our places, these Kingdoms,
But no longer at ease here, in the old dispensation,
With an alien people clutching their gods.
I should be glad of another death.[5]

We are early in the birth of the new age and birth is always painful. Yet holding onto our old models of higher education, to accreditation, to preparing people for the old world, we feel unease ". . . in the old dispensation/With an alien people clutching their gods."

Developing Better Humans for the Age of AI

I spend a lot of time in the education-technology sphere—that community of innovators, investors, and educators—thinking about ways to improve, extend, and, for many of those involved, make money from education. Every April, thousands of us travel to San Diego, California, for the ASU + GSV Summit, the Lollapalooza of edtech, a combination of conference, networking, and speed dating for startup founders and investors. It is the event everyone loves to complain about (too crowded, too transactional, too long lines for coffee) and no one wants to miss. For most of those attending, one foundational belief is so evident that it does not require recitation: higher education is broken and ripe for disruption.

Yet for all the talk of disruption in higher education—whether around for-profit providers, massive open online courses, coding bootcamps, or competency-based education—higher education has been pretty resilient and has still not experienced the kind of off-the-cliff disruption that occurred in industries like music, journalism, or retail. Higher education is regulated through the federal financial aid system and plays multiple roles outside of educating students, such as research, sports entertainment, and serving as a local economic engine. Higher education has been largely protected from disruption, but not from criticism. The public, in the form of employers, policymakers, and families, are demanding change, and the fundamentals of the industry, from demographics to business models to regulatory and governance frameworks, are in disarray. If we are in the AI earthquake as a society, higher education is standing on sand, not bedrock. Its resilience and sheer size and variety will buy it time, but it will not guarantee relevance in a world that will look far different than the Industrial Age it was designed to serve.

It is not too early to start planning for the higher education we need when the earthquake is behind us and the aftershocks have ceased. The question this book examines is, "What does postsecondary education look like in the new Golden Age that follows the earthquake?" One cannot answer that without some hypothesis about what the new world order looks like as AI reshapes our society. As argued previously, in terms of workforce, I believe we will see massive displacement of white-collar jobs, many of them high-paying and high-status roles in finance, technology, insurance, accounting, consulting, law, and more. Will the jobs in those industries go away entirely? No, but we will need far fewer accountants, analysts, software engineers, and others as AI takes over more of the low- and medium-level work, and eventually in some cases, the high-level work. When talent migrates to care economy jobs, the people performing them will work with AI partners that will feel human in many ways, that will be "smarter" than us in many ways, but that are not fundamentally equipped to do the very human work of nurturing a child's development, of comforting someone in distress, of managing someone having a breakdown.

Earlier chapters unpacked the ways AI can change learning. We will finally be able to deliver precision learning, finely tuned to the needs and context of individual learners, and at radically low cost. Knowledge transfer will no longer be the real value-add of the university or the classroom, meaning we can rethink education, the ways we deliver it, and the roles of the people and the machines involved. In education, as in society, we will learn to live with the robots, and that means our roles will change. The education ecosystem will inevitably change to reflect and accommodate these tectonic shifts. All of this adds up a core argument: we will be able to deliver education in new and more powerful ways to prepare students for a new society and workforce.

What, then, is a university for? The answer is simple and profoundly difficult: the university needs to develop better humans, and by extension, a better society. This brings us back to our starting point: the shift from a primary focus on epistemology to ontology, from "teach me what I need to know" to "help me become a flourishing human being." Yes, students will still come to universities and other educational providers to gain "foundational knowledge and skills," whether through degree programs, micro-credential offerings, or just-in-time learning modules. While that learning can be delivered on smart AI-driven platforms like the one we were developing at Matter and Space, it will become secondary to a more important focus for the university of the future—developing our distinctly human skills and capacities.

It has become almost commonplace to say, "We need to spend more time developing the human skills of our students." But what does that mean and how does it reshape higher education? There are four major shifts that should become the cornerstones of the university the world needs next. They are developing moral imagination, developing our embodied selves, developing our ensouled selves, and developing our interpersonal skills. Together, we might call these our human skills or attributes, the ones that distinguish us from the machine. It is not lost on me that these are mostly secondary areas of interest in academia, if explicitly addressed at all. Yet they may come to define and even save higher education in the future.

Developing Moral Imagination

Think of moral imagination as the ability to identify, understand, and navigate moral and ethical issues. It requires the empathy and ability to understand other perspectives, project forward the implications for various decisions, and creativity and commitment to find solutions that work for all. What might that look like?

- A student designing a self-driving truck algorithm might study historical narratives of industrial automation, the human cost of efficiency, to build a humane product. That student might study Michel Foucault to understand how power is embedded in systems or examine the Roman Empire's reliance on infrastructure to understand the fragility of technological dominance. The student might look at social media platforms and the design decisions that earlier engineers and product teams made to feed addiction (they use the term *engagement*) and limit parental controls and insights, leading to terrible consequences. Their ethical framework must be informed by human narratives, not just code, reminding them all technology has ethical and moral dimensions.

- A nursing student might read narratives of suffering (literature) and the moral dimensions of end-of-life care (ethics). The student might study the history of disease and the human response to it, knowing that AI can provide diagnostic facts, but only human judgment can contextualize the emotional truth, including navigating hard conversations in which a patient's values stand in stark contrast to their own. The student could read fiction that explores the ambiguity of choice and the limits of power, recognizing that a treatment plan is also in some ways a moral document, or at least an intensely human construct.

- A business student might confront the tragedy of the commons (ecology) and the impact of unchecked ambition (drama). The student might wrestle with the narratives of ethical failure—tobacco companies hiding health data or oil companies burying projections on climate impact—to see that the drive to maximize profit can lead to a myopic calculus, but that building lasting value for society requires a moral imagination informed by history and philosophy. If the AI partner can draft a flawless prospectus, the student's job might be to ask, "Should this be done, and for whom, and is what I am proposing here ethical and equitable?"

Much of our higher education sector tends to eschew moral and ethical education, even as it has substituted professional preparation for intellectual development as its highest priority (on that point, Father John Henry Newman, author of 1853's *The Idea of the University*, would disagree with such utilitarian purposes of education).[1] It has not served us well, I might argue, and will prove even more inadequate for the Age of AI.

A skeptic might ask, "But whose moral and ethical frameworks?" Part of developing empathy and moral imagination is to teach the various ways societies have tried to frame those questions, and to find ways for students to observe and experience them in practice, whether through experimental learning, case studies, simulations, or real-world projects. As those examples illustrate, developing moral imagination is cross-disciplinary work, weaving into any discipline the ways of thinking we often associate with the humanities and disciplines like philosophy, literature, religion, or theology. These are disciplines much neglected by today's students, providing less than 10 percent of degrees conferred in the United States and increasingly subject to the budgetary chopping block.[2] But they should become foundational to our future models of learning, not necessarily as stand-alone courses or majors, but as humanities restructured, recalibrated, and newly relevant in a world that looks quite different than ours—one that will need what the humanities has to offer like never before.

Developing Our Embodied Selves

We often come back to distinctions between the "thinking" power of AI, such as pattern recognition, prediction, and generative output, and our own cognitive powers. The most important distinction may be that AI "intelligence" lives in data and algorithms, while ours resides in our physical selves. As such, our intelligence is "embodied." Being embodied means that our human cognition is fundamentally grounded in having a physical body that interacts with the real, dynamic world, processing constant sensory-motor feedback within any environment in which we find ourselves. This enables us to form concepts like gravity and balance through lived physical experience, which contrasts sharply with current AI (like large language models) that learns its "world model" abstractly through statistical patterns in digital data. Our embodied intelligence is wondrous and, in many ways, mysterious—like the way we understand that our gut bacteria influences our psychological state or how our breathing and heart rates synchronize when we are in a meeting or other group activity. This is not a conscious effort, but an unconscious physiological mirroring known as interpersonal physiological coherence. This synchrony has been linked to better collaboration, higher levels of empathy, and more effective communication within a group. AI does not possess that power. What we call a "gut feeling," another example of embodied cognition and wisdom, demonstrates how our high-level judgment is instantly informed by our physiological state. This intuition is rooted in the enteric nervous system, or "second brain," which triggers rapid subconscious signals like stomach clenching or a heart rate shift in response to subtle environmental cues. It was the feeling of growing dread as I watched Grady Little leave Boston Red Sox ace Pedro Martinez on the pitcher's mound during game seven of the American League Championship in 2003. A gut feeling is a powerful synthesis of memory and visceral sensation that enables us to make quick, nonconscious assessments. At that moment, as Little turned to walk away, I just knew we would lose the game (and I was right). While we marvel at the speed of AI's computational prowess in a world of digital information, we should not forget our own high-speed processing for navigating the complexity of the physical world as embodied intelligences.

To develop our students' embodied selves, the university the world needs next will make the arts central to learning, especially dance, music, and theater. In an era of digital alienation, where physical presence is diminished and interpersonal intuition and skills are eroding, the arts demand that students operate *in their bodies, in real time,* and *with others.* As mentioned, there may be a time when AI-powered robots can perform flawless dance or produce perfect music, but the point is not about perfection. The drama and power of human art reside in its imperfection, its struggle, and its intentionality. We do not attend the ballet to see a physically perfect *fouetté*; we attend to see the *risk* of the performance, the unique grace of the individual dancer, and the shared embodied emotion in the room. And in that space between performer and audience is often a third state, the one in which performers draw on our energy and we, in turn, rise as one in applause and gratitude when they finish, often leaving emotionally spent. The arts cultivate the capacity for aesthetic judgment and the appreciation of the non-calculable, the things that make life beautiful, not just efficient, and that often engage our embodied selves.

Developing Our Ensouled Selves

More complex in many ways than the workings of our embodied selves are the working of our *ensouled* selves. To be ensouled implies the possession of a unique inner life, consciousness, and subjective experience—attributes often associated with the soul or spirit, or psyche if one prefers. This is the difference between *simulating* intelligence and *being* sentient; while an AI can generate text that expresses compassion, it lacks the genuine subjective emotional experience that underpins true empathy, granting humans unique moral status and the capacity for judgment and ethical deliberation. In the end, AI may perform compassion or emotional connection, but it does not have an authentic emotional life; it does not care if you die or live, are sad or happy, or whether we exist as a species, for that matter (feeding the fears of AI doomsayers).

Our ensouled selves enrich our existence well beyond the limits of the intelligence that AI is starting to mimic so well. Here again, the arts have a role in nurturing this part of our very human "intelligence." When we are moved to tears by a piece of music, a painting, or a poem, we are having a subjective conscious emotional response to perceived beauty, as can happen in nature. I traveled to Antarctica and found myself overwhelmed by the sublime scale and fierce beauty of that environment. At one point, I was sitting alone in the ship's library and started to weep, not out of sadness or joy, but something more subtle and powerful, what aesthetics has sometimes called a sense of "the sublime." Someone later asked if I wept in recognition of my insignificance in the universe, but it was quite the opposite. I wept at my sense of *connection* to something much bigger, a kind of transcendental experience that many would call religious or mystical. It echoed the deeply connected awe I have experienced while being with someone as they took their last breath, as with my mother, or their first breath in this world, as with my daughters. I felt, from that first second I saw my baby girls, the kind of unconditional love that my mother felt for me. It is our ensouled self that processes that which surpasses rational understanding or algorithmic processing. The university of the future will find ways to develop the ensouled intelligences of its students.

Developing Our Interpersonal Skills

As is true today, the university will be an intentional community, but one designed to foster developing the interpersonal, such as collaboration, active listening, communication, empathy, navigating other perspectives and value systems, and more. With knowledge transfer so highly personalized through AI, the need for common experiences and interactions with others will be critical in developing the human skills we will need in a care economy and creating an inclusive, equitable, and just society. While AI may know what we *want* to learn and should learn in terms of foundational knowledge and skills, the human teacher knows what we *need* to learn, which often involves confrontation, ambiguity, and discomfort of the "other." The classroom, the studio, and the residential hall must become sites of intentional *friction* where students confront perspectives that the AI-powered algorithm would never recommend and navigate the messiness of human irrationality, emotion, and biases, which AI would struggle to re-create.

Imagine that our genius teaching assistant has summarized all the legal and historical precedents for a constitutional issue. In the learning studio, the professor assigns students to argue positions contrary to their deeply held political beliefs, forcing them to understand the *logic* of the opposition. The focus is not winning, but empathy for the opposition's perspective, a skill sorely lacking in our polarized digital sphere. The human mentor's intervention is crucial here, stopping the debate not when there is a winner, but when there is a genuine moment of mutual understanding. Or imagine a hypothetical situation that includes a multidisciplinary team of students: a computer science major (building the AI), a psychology major (analyzing behavioral impact), and a public policy major (writing the legislation). They are tasked with addressing a simulated AI loan-default prediction system that shows racial bias. The instructor guides the process, forcing the coders to grapple with the sociological implications of their algorithms, and the policy students to understand the mathematical roots of the injustice. The output of learning becomes a critique of the process, not just an improved product. Students learn not just to use the tool, but to *govern* it. This experience in navigating ambiguity and trade-offs, where there is no clean, clear answer, is the essence of practical wisdom.

These four cornerstone areas, all rooted in human experience and capacity, define education for the Age of AI, while distinguishing us from the machine. Far from diminishing the role of the university or education, this model reclaims higher education's highest purpose: helping students and society reach their fullest potential. This new model of education is in many ways a very old model, harkening back to the very roots of the university in medieval times and on the principle of universal knowledge, designed to create citizens with the intellectual capacity to master any field, and the wisdom and judgment to work across fields or disciplines. It evokes even earlier times when Plato and Aristotle liked the pursuit of knowledge to "the good life" and critical to creating a "virtuous society." We will be asking more, not less, of our universities. That includes guiding students toward self-authorship, the ability to choose one's purpose, to define one's meaning, and to commit to action. The shift from asking, "What do I need to *know*?" to "What should I *become*?" demands a different university than what we mostly see today. It moves the students from a fixed career path to a perpetual, existential quest for purpose within a moral and ethical framework. Students entering the university today, assailed by anxiety, digital isolation, and the gnawing question of whether their careers will be automated, are desperately seeking two things: a good job (the epistemological promise) and a meaningful life (the ontological truth). The university's reinvention must satisfy both, prioritizing the latter as the enduring goal. It must teach students that their dignity and happiness reside not in their utility to the market, but in their capacity for relationship, love, and judgment. The final measure of a great university will not be its endowment size or its research output, but the moral orientation and capacity for flourishing it instills in its graduates. Did the student learn to think critically? Yes. But more important: Did they learn to listen across difference? Did they learn to lead with compassion? Did they learn that the most profound acts of creation and care are those that machines cannot touch?

The university's purpose is not to compete with the machine at its own game—speed, scale, calculation, and data recollection—but to transcend it by dedicating itself fully to what the machine can never replicate: wisdom, judgment, and the capacity for love, empathy, and meaning. The challenge of the Age of AI is not one of technology, but one of will and culture. It is an invitation really, an invitation to make the future humane and the university, with its unique place in society, once again the institution best positioned to teach us how. The alternatives are not great. Our political leaders? Our corporations?

With AI, we have created an intelligence that is, in many ways, greater than our own. The lessons of Mary Shelley's *Frankenstein* still ring true: the failure to love what we create, to nurture and guide its *coexistence* with us with moral and compassionate intentionality, is what makes the monster. AI is not a monster; it is a powerful force that requires us to be more human, more intentional, and more courageous than we have ever been. While AI is perhaps an intelligence greater than our own in terms of our left-brain capacity for pattern matching, rule-bound logic, symbol systems, and reasoning, it fails miserably at what the right brain does so well. As Christian Wiman summarizes the argument of Iain McGilchrist's work on brain hemispheres:

The right brain sees in wholes (the gestalt), whereas the left brain loves systems. The right brain knows what it doesn't know. It's the source of intuition and transformative leaps in all disciplines, including math and science. For the left brain, anything outside its purview is irrelevant, wrong, or invisible. The right brain imagines; the left brain analyzes. The right brain produces (and understands) metaphor; the left brain is more rigidly literal. Poetry comes from the right brain, but, interestingly, language largely comes from the left. And that right there is the key to understanding our divided brains: though we speak of their different capacities, in fact, the left and right are indissolubly linked and can't function healthily without each other. But this health—individual and cultural—depends upon the right brain, which is larger, being the master, and the left brain being the emissary. We have reversed the order.[3]

AI extends, supercharges our left brain capabilities, but it is in development of our right brain capacities that we not only hold on to what differentiates us as humans, but in which we display a superiority that AI cannot begin to approach, a superiority not limited, but expanded by our embodied and ensouled existences in ways that we have scarcely begun to understand.

The revolution is not in the algorithm; it is in our very choices. And what we have argued for in this book and particularly in this chapter may feel revolutionary. We stand at the apex of Carlota Perez's turbulent "installation period" where the benefits are stunning, but the social costs are nearly unbearable. We can choose to cling to the broken models of the past, with the resulting inequities, social divisions, and increased dangers to our health, our planet, our democracy, our children and grandchildren. Or we can embrace the transition with integrity and imagination, using the new abundance generated by AI to fund human infrastructure, universalize access, and make the cultivation of wisdom the central, indispensable purpose of the university. This is the only viable path forward. It is an invitation to faculty, administrators, and policymakers alike to abandon the defense of the past and become the architects of a profoundly human future.

It is *precisely* because we are so broken as a society that radical change is possible. As Perez once commented to me, it is far easier to be bold and build anew *after* the earthquake. We must survive the current time and the troubles ahead. We may have to watch as things get worse before we see the population rise up, through civil unrest and the ballot box, to demand the kinds of policies, controls, regulatory actions, and wealth redistribution required to make the transition period less painful for all. For the ultimately optimistic vision of what *can be* offered in this book, we cannot gloss over the difficulties and dangers that lie ahead. Nor should we delay in thinking through what is required of higher education in these difficult times and in the new world order that will eventually emerge.

The vision outlined in this chapter sounds utopian, I know. But is it utopian to argue for a model of higher education that helps humans and society flourish? To argue for education as a human right? To make central to education developing the now derisively labeled "soft skills" of empathy, collaboration, communication, creativity, and ethical reasoning? To move from an educational model that celebrates individual achievement in an increasingly Darwinian world of winners and losers to one that fosters community and societal well-being? Perhaps it is utopian, but what we have in place today is not working. We are losing the fight against climate change. We are losing the fight against mental illness and addiction. We are losing the fight against authoritarianism. We are losing the fight against rapacious corporations and their billionaire owners, busy building bunkers for the dystopia they are helping create. AI might fully usher in the dystopian future, as many fear. But human intelligence is not doing all that well right now, and AI might be the forcing factor in cementing the death of a failing old world order, while giving us a powerful new tool for building the new world order we so desperately need.

Notes

Chapter 1

1. Ong, Walter J., *Orality and Literacy: The Technologizing of the Word* (Routledge, 1982) 33.
2. Goody, Jack, *The Domestication of the Savage Mind* (Cambridge University Press, 1977).
3. Havelock, Eric, "The Greek Legacy," in David Crowley and Paul Heyer, *Communication in History: Technology Culture, Society*, Third Edition (Longman, 1999) 54–60.
4. Hu, Crystal, "ChatGPT Sets Record for Fastest-Growing User Base— Analyst Note," Reuters, February 2, 2023, https://www.reuters.com/technology/chatgpt-sets-record-fastest-growing-user-base-analyst-note-2023-02-01/
5. Agrawal, Ajay, Avi Goldfarb, and Joshua Gans, *Power and Prediction: The Disruptive Economics of Artificial Intelligence* (Harvard Business Review Press, 2022).
6. Perez, Carlota, *Technological Revolutions and Financial Capital: The Dynamics of Bubbles and Golden Ages* (Edward Elgar Publishing, 2003).
7. Singer, Natasha, "Goodbye, $165,000 Tech Jobs. Student Coders Seek Work at Chipotle," *New York Times*, August 10, 2025, https://www.nytimes.com/2025/08/10/technology/coding-ai-jobs-students.html
8. Perez, *Technological Revolutions and Financial Capital*.
9. Perez, *Technological Revolutions and Financial Capital*.
10. David Forgacs, ed. *The Antonio Gramsci Reader: Selected Writings 1916–1935* (New York University Press, 2000).

Chapter 2

1. Turing, Alan M., "Computing Machinery and Intelligence," *Mind* LIX (236) October 1950, 433–460, https://doi.org/10.1093/mind/LIX.236.433

2. McCarthy, John, et al., "A Proposal for the Dartmouth Summer Research Project on Artificial Intelligence," August 31, 1955, https://www-formal.stanford.edu/jmc/history/dartmouth/dartmouth.html

3. "Artifacts from the Future," The Institute for the Future, https://legacy.iftf.org/what-we-do/artifacts-from-the-future/

4. Kaye, Danielle, "Nvidia Hits New Milestone as World's First $5tn Company," BBC, October 29, 2025, https://www.bbc.com/news/articles/cp8e970vn5vo

5. Weber, Bruce, "Swift and Slashing, Computer Topples Kasparov," *New York Times*, May 12, 1997, https://www.nytimes.com/1997/05/12/nyregion/swift-and-slashing-computer-topples-kasparov.html

6. Vaswani, Ashish, et al., "Attention Is All You Need," *NIPS'17: Proceedings of the 31st International Conference on Neural Information Processing Systems*, December 4, 2017, 6000–6010, https://dl.acm.org/doi/10.5555/3295222.3295349

7. Perez, Carlota, "What Is AI's Place in History?" Project Syndicate, March 11, 2024, https://www.project-syndicate.org/magazine/ai-is-part-of-larger-technological-revolution-by-carlota-perez-1-2024-03

8. Rijo, Luis, "LeCun Calls Auto-Regressive LLMs "Doomed" at NYU Seminar," PPC Land, September 25, 2025, https://the-decoder.com/the-case-against-predicting-tokens-to-build-agi/

Chapter 3

1. Townsend, Robert B., and Norman Bradburn, "The State of the Humanities Circa 2022," *Daedalus* 151 (3) 2022, 11–18, https://direct.mit.edu/daed/article/151/3/11/112685/The-State-of-the-Humanities-circa-2022

2. "PISA 2022 Assessment and Analytical Framework," Organisation for Economic Co-operation and Development, August 31, 2023, https://www.oecd.org/en/publications/pisa-2022-assessment-and-analytical-framework_dfe0bf9c-en/full-report.html

3. Yee, Lareina, et al., "Agents, Robots, and Us: Skill Partnerships in the Age of AI," McKinsey Global Institute, November 25, 2025, https://www.mckinsey.com/mgi/our-research/agents-robots-and-us-skill-partnerships-in-the-age-of-ai

4. Sigelman, Matt, et al., "Shifting Skills, Moving Targets, and Remaking the Workforce," Burning Glass Institute, May 2022, https://static1 .squarespace.com/static/6197797102be715f55c0e0a1/t/628676792 e42956192887397/1652979369435/Shifting-Skills-Moving-Targets-and-Remaking-the-Workforce+BGI+and+BCG.final.pdf

5. Darley, James, "'White-Collar Bloodbath': Anthropic Warns of AI Job Losses," *Technology*, June 9, 2025, https://technologymagazine.com/ articles/white-collar-bloodbath-anthropic-warns-of-ai-job-losses

6. "The Generative AI Adoption Tracker," Project on the Workforce, November 13, 2025, https://pw.hks.harvard.edu/post/the-generative-ai-adoption-tracker

7. Apotheker, Jessica, et al., "The Widening AI Value Gap: Build for the Future 2025," Boston Consulting Group, September 2025, https:// media-publications.bcg.com/The-Widening-AI-Value-Gap-Sept-2025.pdf

8. Wangman, Ryan, "Everyone Agrees AI Is Transformative. Whether That's A Plus Is a Debate for Architecture and Design Pros," *Bisnow Chicago Architecture & Design*, April 15, 2025, https://www.bisnow.com/ chicago/news/architecture-design/chicago-architecture-and-design-pros-divided-over-future-of-ai-in-creative-work-128912

9. Shein, Esther, "Anthropic CEO: AI Will Soon Take Nearly Half of Entry-Level White-Collar Jobs," eWeek, May 29, 2025, https://www.eweek.com/ news/anthropic-dario-amodei-ai-white-collar-jobs/

10. Burleigh, Emma, "Google DeepMind CEO Demis Hassabis Says That Humans Have Just over 5 Years Before AI Will Outsmart Them," *Fortune*, March 18, 2025, https://fortune.com/2025/03/18/google-deepmind-ceo-demis-hassabis-agi-outsmart-human-workers-job-replacement-ai/

11. Suleyman, Mustafa, *The Coming Wave: AI, Power, and Our Future* (Crown, 2025).

12. "Deepfake Robocall in New Hampshire Brings AI-Fueled Disinformation to Election 2024," The Center for Election Innovation & Research, January 2024, https://electioninnovation.org/update/deepfake-robocall-in-new-hampshire-brings-ai-fueled-disinformation-to-election-2024/

13. Russell, Stuart, *Human Compatible: Artificial Intelligence and the Problem of Control* (Penguin Books, 2020).

14. Piketty, Thomas, *Capital in the Twenty-First Century* (Belknap Press, 2014); Stiglitz, Joseph, *The Price of Inequality: How Today's Divided Society Endangers Our Future* (W. W. Norton & Company, 2013).

15. Gilens, Martin, and Benjamin Page, "Testing Theories of American Politics: Elites, Interest Groups, and Average Citizens," *Perspectives on Politics* 12 (3) September 18, 2014, 564–581, https://www.cambridge.org/ core/journals/perspectives-on-politics/article/abs/testing-theories-of-american-politics-elites-interest-groups-and-average-citizens/ 62327F513959D0A304D4893B382B992B

16. Perez, Carlota, *Technological Revolutions and Financial Capital: The Dynamics of Bubbles and Golden Ages* (Edward Elgar Publishing, 2003).

17. Delaney, Kevin J., "Bill Gates: This Is Why We Should Tax Robots," World Economic Forum, February 20, 2017, https://www.weforum.org/stories/2017/02/bill-gates-this-is-why-we-should-tax-robots/

18. Perez, *Technological Revolutions and Financial Capital.*

19. Autor, David H., "Why Are There Still So Many Jobs? The History and Future of Workplace Automation," *Journal of Economic Perspectives* 29 (3) Summer 2015, 3–30, https://pubs.aeaweb.org/doi/pdfplus/10.1257/jep.29.3.3

20. Putnam, Robert D., *Bowling Alone: The Collapse and Revival of American Community* (Simon & Schuster, 2000).

21. "The Science of Well-Being," https://www.drlauriesantos.com/science-well-being

Chapter 4

1. "Nearly 3 in 4 Teens Have Used AI Companions, New National Survey Finds," Common Sense Media, July 16, 2025, https://www.commonsensemedia.org/press-releases/nearly-3-in-4-teens-have-used-ai-companions-new-national-survey-finds

2. Nicholas, Epley, Waytz, Adam, and Cacioppo, John, "On Seeing Human: A Three-Factor Theory of Anthropomorphism," *Psychological Review* 114 (4) 2007, 864–886, https://pubmed.ncbi.nlm.nih.gov/17907867/

3. Metzinger, Thomas K., "Empirical Perspectives from the Self-Model Theory of Subjectivity: A Brief Summary with Examples." *Progress in Brain Research* 168 2008, 215–245, https://www.semanticscholar.org/paper/Empirical-perspectives-from-the-self-model-theory-a-Metzinger/8fba3f942096ce434cff9ffb4e39f8fb56912383

4. Turkle, Sherry, *Alone Together: Why We Expect More from Technology and Less from Each Other* (Basic Books, 2012).

5. Harari, Yuval Noah, *Home Deus: A Brief History of Tomorrow* (Harper, 2017).

6. Dreyfus, Hubert, and Stuart E. Dreyfus, *Mind Over Machine: The Power of Human Intuition and Expertise in the Era of the Computer* (Free Press, 1988).

7. Warr, Melissa, et al., "A Chat About GPT3 (and Other Forms of Alien Intelligence) with Chris Dede," *TechTrends* 67 2003, 396–401, https://link.springer.com/article/10.1007/s11528-023-00843-z

8. Goldstein, Pavel, et al., "Brain-to-Brain Coupling During Handholding Is Associated with Pain Reduction," *Proceedings of the National Academy of Sciences of the United States of America*, February 26, 2018, https://www.pnas.org/doi/full/10.1073/pnas.1703643115

9. Crawford, Kate, and Trevor Paglen, "Excavating AI: The Politics of Images in Machine Learning Training Sets," The AI Now Institute, September 19, 2019, https://excavating.ai/

10. Nass, Clifford, et al., "Can Computer Personalities Be Human Personalities?" *International Journal of Human-Computer Studies* 43 (2) 1995, 223–239, https://www.sciencedirect.com/science/article/abs/pii/S1071581985710427

11. Benjamin, Ruha, *Race After Technology: Abolitionist Tools for the New Jim Code* (Polity, 2019).

12. Epley, Nicholas, Adam Waytz, and John T. Cacioppo, "On Seeing Human: A Three-Factor Theory of Anthropomorphism," *Psychological Review* 114 (4) 2007, 864–886, https://pubmed.ncbi.nlm.nih.gov/17907867/

13. Turkle, *Alone Together*.

14. Friere, Paulo, *Pedagogy of the Oppressed* (Bloomsbury Academic, 2018).

15. hooks, bell, *Teaching to Transgress: Education as the Practice of Freedom* (Routledge, 1994).

16. Papert, Seymour, *Mindstorms: Children, Computers, and Powerful Ideas* (Basic Books, Inc., 1982).

17. Harari, *Home Deus*.

18. Ming, Lee Chong, and Kelsey Vlamis, "The Godfather of AI Has a Tip for Surviving the Age of AI: Train It to Act Like Your Mom," *Business Insider*, August 14, 2025, https://www.businessinsider.com/godfather-of-ai-maternal-instincts-humanity-survival-geoffrey-hinton-2025-8

19. Greenfield, Beth, "Nearly Half of Gen Zers Wish TikTok 'Was Never Invented,' Survey Finds," *Fortune*, September 17, 2024, https://fortune.com/well/article/nearly-half-of-gen-zers-wish-social-media-never-invented/

20. India, Freya, "Time to Refuse," After Babel, October 5, 2025, https://www.afterbabel.com/p/time-to-refuse

Chapter 5

1. Young, Jeffrey R., "Some Experts Skeptical of AI Tutors, Recalling IBM's 'Watson'," *Government Technology*, January 22, 2024, https://www.govtech.com/education/higher-ed/some-experts-skeptical-of-ai-tutors-recalling-ibms-watson

Chapter 6

1. Reed, Jon, "OpenAI Pulled a Big ChatGPT Update. Why It's Changing How It Tests Models," CNET, May 5, 2025, https://www.cnet.com/tech/services-and-software/openai-pulled-a-big-chatgpt-update-why-its-changing-how-it-tests-models/

2. Milmo, Dan, and Alex Hern, "'We Definitely Messed Up': Why Did Google AI Tool Make Offensive Historical Images?" *The Guardian*, March 8, 2024; Sherr, Ian, "Perplexity AI Results Include Plagiarism and Made-Up Content, Reports Say," CNET, June 20, 2024, https://www.cnet.com/tech/services-and-software/perplexity-ai-results-include-plagiarism-and-made-up-content-reports-say/; Lee, Dave, "Tay: Microsoft Issues Apology over Racist Chatbot Fiasco," *BBC*, March 25, 2016, https://www.bbc.com/news/technology-35902104

3. Legatt, Ava, "90% Of College Students Use AI: Higher Ed Needs AI Fluency Support Now," *Forbes*, September 18, 2025, https://www.forbes.com/sites/avivalegatt/2025/09/18/90-of-college-students-use-ai-higher-ed-needs-ai-fluency-support-now/

4. "What Students Want: Key Results from DEC Global AI Student Survey 2024," Digital Education Council, August 7, 2024, https://www.digitaleducationcouncil.com/post/what-students-want-key-results-from-dec-global-ai-student-survey-2024

5. McKerin, Cheryl, "Report on Student Attitudes Towards AI in Academia," University of Illinois Chicago, April 5, 2024, https://learning.uic.edu/news-stories/report-on-student-attitudes-towards-ai-in-academia/

6. Full disclosure: I was a reviewer for the publication. "Student Guide to Artificial Intelligence 2025," American Association of Colleges and Universities and Elon University, https://www.aacu.org/publication/student-guide-to-artificial-intelligence

7. Burton, Graeme, "AI-Led Automation Is Already Cutting Entry-Level Jobs, BSI," *Computing*, October 9, 2025, https://www.computing.co.uk/news/2025/ai/ai-led-automation-cutting-entry-level-jobs

8. "Evolving Together: AI, Automation and Building the Skilled Workforce of the Future," BSI Group, 2025, https://www.bsigroup.com/siteassets/pdf/en/insights-and-media/insights/white-papers/flourishing-in-the-ai-workforce.pdf?utm_source=onedtech.philhillaa.com&utm_medium=newsletter&utm_campaign=interesting-reads-this-week&_bhlid=f6bee0921236941858336142a28d4587b15a58df

9. Jenay, Robert, "2024 EDUCAUSE AI Landscape Study," EDUCAUSE, February 26, 2024, https://www.educause.edu/ecar/research-publications/2024/2024-educause-ai-landscape-study/policies-and-procedures

10. "Future of Jobs Report 2023," World Economic Forum, May 2023, https://www3.weforum.org/docs/WEF_Future_of_Jobs_2023.pdf

11. Ellingrud, Kweilin, et al., "Generative AI and the Future of Work in America," McKinsey Global Institute, July 26, 2023, https://www.mckinsey .com/mgi/our-research/generative-ai-and-the-future-of-work-in-america

12. Frey, Carl Benedikt, and Michael Osborne, "The Future of Employment," Oxford Martin Programme on Technology and Employment, September 17, 2013, https://oms-www.files.svdcdn.com/production/downloads/academic/future-of-employment.pdf

13. Newman, John Henry, *The Idea of a University* (Assumption Press, 2014).

14. Newman, *The Idea of a University*.

15. Perez, Carlota, and Tamsin Murray Leach, "A Smart Green 'European Way of Life': The Path for Growth, Jobs and Wellbeing," (Working paper) Beyond the Technological Revolution, March 2018, https://carlotaperez.org/wp-content/downloads/new-book/outputs/working-papers/BTTR_WP_2018-1.pdf

Chapter 7

1. "Gregory Elliott," Brown University Department of Sociology, https://sociology.brown.edu/people/gregory-elliott

2. Beegle, Donna M., *See Poverty . . . Be the Difference* (Communication Across Barriers, 2007).

3. Keyes, Corey L. M., "Risk and Resilience in Human Development: An Introduction," *Research in Human Development* 1 (4) 2004, 223-227, https://www.researchgate.net/publication/241713828_Risk_and_Resilience_in_Human_Development_An_Introduction

4. LeBlanc, Paul J., *Broken: How Our Systems of Care Are Failing Us and How We Can Fix Them* (Matt Holt Books, 2022).

5. Mazur, Eric, *Peer Instruction: A User's Manual* (Pearson, 1996).

6. This is another example of Chris Dede's distinction between wisdom or judgment and the predictive, information processing power of AI.

Chapter 8

1. LeBlanc, Paul J., *Students First: Equity, Access, and Opportunity in Higher Education* (Harvard Education Press, 2021).

2. Full disclosure: I serve on the board of a proposed outcomes-focused accreditor, the Postsecondary Commission.

3. Lange, Jason, and Alexandra Alper, "Americans Fear AI Permanently Displacing Workers, Reuters/Ipsos Poll Finds," Reuters, August 20, 2025, https://www.reuters.com/world/us/americans-fear-ai-permanently-displacing-workers-reutersipsos-poll-finds-2025-08-19/

4. Perez, Carlota and Tamsin Murray Leach, "A Smart Green 'European Way of Life': The Path for Growth, Jobs and Wellbeing," (Working paper) Beyond the Technological Revolution, March 2018, https://carlotaperez.org/wp-content/downloads/new-book/outputs/working-papers/BTTR_WP_2018-1.pdf
5. Eliot, T. S., "Journey of the Magi." This poem is in the public domain. Published in Poem-a-Day on December 24, 2023, by the American Academy of Poets.

Chapter 9

1. Newman, John Henry, *The Idea of a University* (Assumption Press, 2014).
2. "The Number of College Graduates in the Humanities Drops for the Eighth Consecutive Year," American Academy of Arts & Sciences, November 22, 2021, https://www.amacad.org/news/college-graduates-humanities-drops-eighth-consecutive-year
3. Wiman, Christian, "The Tune of Things: Is Consciousness God?" *Harper's Magazine*, December 2025, https://harpers.org/archive/2025/12/the-tune-of-things-christian-wiman-consciousness-god/

Acknowledgments

Like the narrator of the T. S. Eliot poem included at the end of chapter eight, I wrote this book knowing that the old world order is broken and falling away, that the new world order is not yet clearly defined, and that the transition will be messy, painful, and hard to navigate. When Ryan Flahive, my acquisitions editor at Wiley, asked me to write the book, I found the prospect daunting since AI, as we think of it today, is still so new and evolving so quickly. Yet I agreed to do it, because I always turn to writing to make sense of things. So I wrote this book not so much as an expert (who really is at this moment?) but as a learner.

What great teachers I have had on this journey, starting with my coauthors, George Siemens and Tanya Gamby. Even though I took the lead on writing and revising the book, they contributed key parts of it, and, more important, their thinking is everywhere apparent. We have learned together. We also learned a lot from the Matter and Space team, especially Rachel Koblic, Damien Coyle, Ann Wang, Rob English, Matt Beil, Kemp Battle, John Fillmore, Bill Nash, Srecko Joksimovic, and the others at C3L. The larger framing of this historical moment borrows heavily from the work of the amazing Carlota Perez. Imagine my joy when a mutual friend introduced me to Carlota, whom I have long admired. She has been generous in sharing her thinking with me, and my optimism about a new Golden Age, balanced with a sober fear of the difficult transition still ahead of us, very much echoes her sentiments and thinking.

I am grateful to the many others who contributed to this book in ways great and small. They include Patrick Methvin, Ludo Fourrage, and Dale Whittaker from the postsecondary team at the Gates Foundation, as well as Bill G. himself, who in an extraordinary dinner at his home gave me better questions than the ones I had in hand, which is the best of all gifts. Ben Gomes, who heads higher education at Google, and Mike Fleming, a key member of Ben's team, brought me into important discussions. Organizations with which I am involved, including Ellucian, Carnegie Higher Education, and K16 solutions, as well as a host of universities, all contributed to my education as well.

I have a growing list of things for which I need to thank my old friend Ted Mitchell. When he was under secretary of education, he made an opportunity available to me that changed my professional life. In his role leading the American Council on Education, we worked closely together, especially though the pandemic, when I was board chair. Now he has written the foreword to this book, so another item to add to my growing list. Most important, I'm grateful for his friendship.

As a writer, there is no better friend than a great editor. Here I thank Margaret Moffett, with whom I have now collaborated three times. There is no better editor and lest I recall too much Eliot's backhanded compliment of Ezra Pound (if you know, you know), she is also a damn good writer. She makes anything I write better and if I have any other books in me, I will want her at my side when I write them. I'm very grateful to the team at Wiley, especially the aforementioned Ryan and the others who brought the book to life.

Nothing I do on the page, or in life generally, is without the influence of my family. Pat has put up with me as her husband for decades now, celebrating my victories with exuberant pride, holding my grudges with ferocity, and always providing an unerring B. S. meter. As an astute friend says of Pat, "She is the least f**ked-up person I know." High praise. My daughters, Emma and Hannah, are fine writers. Emma is a natural storyteller, on the page, on film, and now in a courtroom. Hannah, also an educator, will have a book out in 2026 as well; it will be better written. The person most present for me when I was writing is Simon, our grandson, who is 18 months old. If we are to eventually enter a new "Golden Age," as Carlota's work predicts, it will be his world, and I ache to have it be one in which he can flourish. He was born into a world terribly broken and with whatever number of years I have left in it, I will do all I can to realize the vision for this book. To create the world that Simon and everyone else's babies deserve.

About the Author

Dr. Paul J. LeBlanc is a visiting scholar and special advisor at the Harvard University Graduate School of Education. He is also president emeritus of Southern New Hampshire University (SNHU). Under the 21 years of Paul's direction, SNHU grew from 2,800 students to over 250,000 and is now the largest non-profit provider of online higher education in the country.

Paul is considered one of America's most innovative educators. *Forbes Magazine* has listed him as one of its 15 "Classroom Revolutionaries" and *Washington Monthly* named him one of America's 10 most innovative university presidents. In 2018, Paul won the prestigious IAA Institute Hesburgh Award for Leadership Excellence in Higher Education, joining some of the most respected university and college presidents in American higher education. In 2024, the University of Pennsylvania awarded him the Zemsky Medal for Innovation in Higher Education. He has also received the Ernest L. Boyer Award (NACU), the Distinguished Alumnus Award (AASCU), and the Ray Schroeder Leadership Award (UPCEA).

He is the author of *Students First: Equity, Access, and Opportunity in Higher Education* (2021), winner of the 2022 Phillip E. Frandson Award for Literature, and *Broken: How Our Social Systems Are Failing Us and How We Can Fix Them* (2022). His wife, Pat, is an attorney, as is his daughter Emma. His other daughter, Hannah, is a high school history teacher and mother to his grandson, Simon, to whom this book is dedicated.

Index